Test Bank for

Principles of Human Anatomy

Eighth Edition

Gerard J. Tortora
Bergen Community College

Janice M. Meeking
Mount Royal College
Calgary, Alberta

An imprint of Addison Wesley Longman, Inc.

Don Mills, Ontario ■ Sydney ■ Mexico City ■ Madrid ■ Amsterdam
Menlo Park, California ■ Reading, Massachusetts ■ New York
Bonn ■ Paris ■ Milan ■ Singapore ■ Tokyo ■ Seoul ■ Taipei

ISBN 0-321-03656-5

1 2 3 4 5 6 7 8 9 10–VG–03 02 01 00 99 98

Benjamin/Cummings Publishing
2725 Sand Hill Road
Menlo Park, California 94025

Preface

This test bank is designed to accompany Gerard J. Tortora's *Principles of Human Anatomy*, Eighth Edition. There are a total of 1775 questions, with up to 85, but no less than 50, questions in each chapter. Question formats in each chapter include multiple choice, true/false, short answer, and essay.

Although this printed test bank can be used alone, the maximum benefit and flexibility are afforded the instructor if the test bank is used with the *Test Gen-EQ* program. *Test Gen-EQ*, which was used to create the test bank, comes with an easy-to-follow comprehensive manual and is available from Addison Wesley Longman (ISBN 0-321-03657-3 for Macintosh; 0-321-03658-1 for Windows). This user-friendly test generator enables the instructors to view and edit existing test bank questions and to easily add questions to customize the test bank to meet their own course requirements. The built-in question editor allows instructors to create questions in six different formats and to import graphics. Multiple views and on-screen tools allow you to choose and transfer questions to tests and print tests in a variety of fonts and forms. The test bank thus provides you with a ready-to-use powerful instrument for creating a variety of classroom tests, ranging from short, subject-specific quizzes to final comprehensive examinations.

Included with the computer test bank is the *Quizmaster EQ* program, which enables instructors to create and save tests using *Test Gen-EQ* so that students can take them on a computer network. Instructors can set preferences for how and when tests are administered. The program automatically grades the exams and allows the instructor to view or print a variety of reports for individual students, classes, or courses. Technical support is available upon request to help you successfully implement the *Quizmaster EQ* course management software.

Additional resources are available to instructors using Gerard J. Tortora's *Principles of Human Anatomy*, Eighth Edition. To find out more, please visit the Tortora web site at **www.awl.com/bc.**

Comments and suggestions on this test bank are welcome and may be sent to the author c/o the Tortora A&P Book Team, Addison Wesley Longman, 2725 Sand Hill Road, Menlo Park, CA 94025, or via email at Kueno@awl.com.

Janice M. Meeking
Mount Royal College
Calgary, Alberta

Contents

CHAPTER 1   An Introduction to the Human Body

MULTIPLE CHOICE.   Choose the one alternative that best completes the statement or answers the question.

1)   The levels of structural organization from least complex to most complex are as follows:
A) chemical, cellular, organ, tissue, system.
B) cellular, organ, chemical, tissue, system.
C) chemical, cellular, tissue, organ, system.
D) chemical, system, tissue, cellular, organ.

Answer: C
Type: MC   Page Ref: 2-5
Topic: LEVELS OF STRUCTURAL ORGANIZATION

2)   A course of study in which you learned about the bones of the forearm, along with associated blood vessels, lymphatic vessels and nerves, would most appropriately be called:
A) gross anatomy.   B) regional anatomy.
C) systemic anatomy.   D) surface anatomy.

Answer: B
Type: MC   Page Ref: 2
Topic: ANATOMY

3)   The four basic types of tissues in the body are:
A) skeletal, muscular, epithelial, nervous.
B) connective, muscle, nervous, epithelial.
C) vascular, nervous, epithelial, connective.
D) muscle, nervous, skeletal, connective.

Answer: B
Type: MC   Page Ref: 4
Topic: LEVELS OF STRUCTURAL ORGANIZATION

4)   Which of the following statements about the urinary system is *not* true?
A) It helps maintain acid-base balance.
B) It secretes a hormone that regulates red blood cell production.
C) It includes kidneys, spleen, ureters, urethra and urinary bladder.
D) It eliminates wastes.

Answer: C
Type: MC   Page Ref: 8
Topic: SYSTEMS: Organs and Functions

5)   Catabolism is an important function of the _____ system.
A) respiratory   B) digestive   C) urinary   D) cardiovascular

Answer: B
Type: MC   Page Ref: 6,8
Topic: LIFE PROCESSES

6)   Liver cells (hepatocytes) are responsible for many important processes; one of which is synthesis of plasma proteins. Production of these proteins would be an example of:
A) the anabolic phase of metabolism.   B) the catabolic phase of metabolism.
C) differentiation.   D) responsiveness.

Answer: A
Type: MC   Page Ref: 6
Topic: LIFE PROCESSES

7)   Which word describes the location of the kidney with respect to the urinary bladder?
A) anterior   B) inferior   C) distal   D) superior

Answer: D
Type: BI   Page Ref: 13
Topic: DIRECTIONAL TERMS

8)   A student pays for one pair of pierced earrings, then has one inserted in the left earlobe and one in the left nostril. The location of the two rings, with respect to each other, may be most accurately described as:
A) contralateral.   B) ipsilateral.   C) superior.   D) intermediate.

Answer: B
Type: MC   Page Ref: 13
Topic: DIRECTIONAL TERMS

9)   In the anatomical position, the palms of the hands are facing:
A) anteriorly   B) posteriorly   C) laterally   D) medially

Answer: A
Type: MC   Page Ref: 9
Topic: ANATOMICAL POSITION

10)   An accident report contains the following statement: "The victim suffered a severe blow to the mental region." The officer making this statement was a former student of anatomy. His statement therefore means:
A) the victim had a severe wound to the skull bones.
B) the victim's brain was obviously injured.
C) the victim had been struck on the chin.
D) the victim had witnessed a shocking event.

Answer: C
Type: MC   Page Ref: 10
Topic: REGIONAL NAMES

11)   A _____ section of the body would reveal the location of the heart with respect to the sternum.
A) transverse   B) coronal
C) oblique   D) any of the above

Answer: A
Type: BI   Page Ref: 12
Topic: PLANES AND SECTIONS

12) A plane that divides the body into superior and inferior portions is:

A) midsagittal.   B) parasagittal.   C) coronal.   D) transverse.

Answer: D
Type: MC   Page Ref: 12
Topic: PLANES AND SECTIONS

13) Serous membranes are associated with the:

A) pleural cavity.   B) pericardial cavity.
C) abdominal cavity.   D) all of the above.

Answer: D
Type: MC   Page Ref: 14
Topic: BODY CAVITIES

14) The pericardial cavity is situated in:

A) the ventral cavity.   B) the thoracic cavity.
C) the mediastinum.   D) all of the above.

Answer: D
Type: MC   Page Ref: 14
Topic: BODY CAVITIES

15) The word "dorsum" may be used to describe the:

A) back of the hand.   B) anterior surface of the body.
C) top of the foot.   D) both A and C.

Answer: D
Type: MC   Page Ref: 10
Topic: REGIONAL NAMES

TRUE/FALSE. Write 'T' if the statement is true and 'F' if the statement is false.

1) Developmental anatomy is the study of anatomical changes from the fertilized egg to the adult.

Answer: FALSE
Type: TF   Page Ref: 2
Topic: ANATOMY

2) Gross anatomy involves the microscopic study of the structure of tissues.

Answer: FALSE
Type: TF   Page Ref: 2
Topic: ANATOMY

3) The cardiovascular system filters body fluids.

Answer: FALSE
Type: TF   Page Ref: 5
Topic: SYSTEMS: Organs and Functions

4) The lymphatic system is responsible for the transportation of oxygen and carbon dioxide between the lungs and body tissues.

Answer: FALSE
Type: TF   Page Ref: 6
Topic: SYSTEMS: Organs and Functions

5) The antebrachial region is distal to the antecubital region.

Answer: TRUE
Type: TF   Page Ref: 10
Topic: REGIONAL NAMES

6) The patellar region is superior to the inguinal region.

Answer: FALSE
Type: TF   Page Ref: 10
Topic: REGIONAL NAMES

7) The knee is proximal to the ankle.

Answer: TRUE
Type: TF   Page Ref: 13
Topic: DIRECTIONAL TERMS

8) Anabolism provides the energy for life processes by breaking down food molecules.

Answer: FALSE
Type: TF   Page Ref: 6
Topic: LIFE PROCESSES

9) The hypogastric region is lateral to the hypochondriac region.

Answer: FALSE
Type: TF   Page Ref: 18
Topic: ABDOMINOPELVIC REGIONS

10) The epigastric region is superior to the hypogastric region.

Answer: TRUE
Type: TF   Page Ref: 18
Topic: ABDOMINOPELVIC REGIONS

11) The ventral body cavity contains the heart and liver.

Answer: FALSE
Type: TF   Page Ref: 13,14
Topic: BODY CAVITIES

12) The serous membrane associated with the lungs is called pleura.

Answer: TRUE
Type: TF   Page Ref: 14
Topic: BODY CAVITIES

13) The descending colon of the large intestine extends from the left lumbar region into the left iliac region.

Answer: TRUE
Type: TF   Page Ref: 18
Topic: ABDOMINOPELVIC REGIONS

ESSAY.   Write your answer in the space provided or on a separate sheet of paper.

1) List the components of the following major systems and give two functions of each system: skeletal, muscular, digestive, respiratory, nervous, urinary.

Answer: Refer to the text and diagrams of Table 1.2, pp. 4–8.
Type: ES   Page Ref: 4–8
Topic: SYSTEMS: Organs and Functions

2) Name and describe six life processes and give an example of each.

Answer: Refer to the text, pp. 6–8.
Type: ES   Page Ref: 6–8
Topic: LIFE PROCESSES

3) Detective I.M. Smart (a former anatomy student) was called to investigate a murder scene. The victim was lying in a supine position, his glazed eyes staring skyward, legs together with toes pointing upwards, arms by his sides, palms facing upward. Smart concluded that the victim was found in anatomical position. Was he correct? Why or why not?

Answer: He was incorrect. In order for the victim's body to be in anatomical position, the body would have to be standing.
Type: ES   Page Ref: 9
Topic: ANATOMICAL POSITION

4) Define the planes of section known as coronal, transverse, midsagittal, and parasagittal. Supplement your answer with an illustration of the three planes in a simple figure.

Answer: coronal: creates anterior and posterior portions
            transverse: creates superior and inferior portions
            midsagittal: creates equal right and left portions
            parasagittal: creates unequal right and left portions
Type: ES   Page Ref: 12
Topic: PLANES AND SECTIONS

5) The abdominopelvic cavity is divided into nine regions. Using two vertical and two horizontal lines, divide your answer space into nine regions and label the abdominopelvic regions.

Answer: See Figure 1.9.
Type: ES   Page Ref: 18
Topic: ABDOMINOPELVIC REGIONS

SHORT ANSWER. Write the word or phrase that best completes each statement or answers the question.

1)   A system consists of related _____ that have a common function.

Answer: organs
Type: SA   Page Ref: 5
Topic: LEVELS OF STRUCTURAL ORGANIZATION

2)   The system responsible for movement of limbs, maintenance of posture, and heat production is the _____ system.

Answer: muscular
Type: SA   Page Ref: 5
Topic: SYSTEMS: Organs and Functions

3)   The _____ system enables the body to detect and respond to environmental change.

Answer: nervous
Type: SA   Page Ref: 6
Topic: SYSTEMS: Organs and Functions

4)   Regulatory chemicals produced by endocrine glands are called _____.

Answer: hormones
Type: SA   Page Ref: 7
Topic: SYSTEMS: Organs and Functions

5)   The right iliac region of the abdomen is _____ to the right lumbar region.

Answer: inferior
Type: SA   Page Ref: 18
Topic: ABDOMINOPELVIC REGIONS

6)   In anatomical position, the thumb is _____ to the index finger.

Answer: lateral
Type: SA   Page Ref: 13
Topic: DIRECTIONAL TERMS

7)   In anatomical position, the great toe is on the _____ side of the foot.

Answer: medial
Type: SA   Page Ref: 13
Topic: DIRECTIONAL TERMS

8)   A section through the body which produces equal right and left portions of an eyeball would be a _____ section of the body.

Answer: parasagittal
Type: SA   Page Ref: 12
Topic: PLANES AND SECTIONS

9)   A section through the left eyeball that produces equal right and left portions of the eyeball would be a _____ section of the eyeball.

Answer: midsagittal
Type: SA   Page Ref: 12
Topic: PLANES AND SECTIONS

10) The left _____ cavity surrounds the left lung.

Answer: pleural
Type: SA   Page Ref: 14
Topic: BODY CAVITIES

11) The two subdivisions of the ventral body cavity are the thoracic and the _____ cavity.

Answer: abdominopelvic
Type: SA   Page Ref: 12
Topic: BODY CAVITIES

12) The right lobe of the liver is located in the _____ region of the abdominopelvic cavity.

Answer: right hypochondriac
Type: SA   Page Ref: 18
Topic: ABDOMINOPELVIC REGIONS

13) The _____ region of the abdominopelvic cavity contains the rectum and the urinary bladder.

Answer: hypogastric
Type: SA   Page Ref: 18
Topic: ABDOMINOPELVIC REGIONS

MATCHING. Choose the item in Column 2 that best matches each item in Column 1.

Match each organ or tissue in Column 1 with its system in Column 2.

1) Column 1: spleen
Column 2: lymphatic

Answer: lymphatic
Type: MA   Page Ref: 6
Topic: SYSTEMS: Organs and Functions

2) Column 1: sweat and oil glands
Column 2: integumentary

Answer: integumentary
Type: MA   Page Ref: 4
Topic: SYSTEMS: Organs and Functions

3) Column 1: jaw bone
Column 2: skeletal

Answer: skeletal
Type: MA   Page Ref: 4
Topic: SYSTEMS: Organs and Functions

4) Column 1: chewing muscles
Column 2: muscular

Answer: muscular
Type: MA   Page Ref: 5
Topic: SYSTEMS: Organs and Functions

5)   Column 1: blood
        Column 2: cardiovascular

        Answer: cardiovascular
        Type: MA   Page Ref: 5
        Topic: SYSTEMS: Organs and Functions

6)   Column 1: spinal cord
        Column 2: nervous

        Answer: nervous
        Type: MA   Page Ref: 6
        Topic: SYSTEMS: Organs and Functions

7)   Column 1: eye
        Column 2: nervous

        Answer: nervous
        Type: MA   Page Ref: 6
        Topic: SYSTEMS: Organs and Functions

8)   Column 1: pituitary gland
        Column 2: endocrine

        Answer: endocrine
        Type: MA   Page Ref: 7
        Topic: SYSTEMS: Organs and Functions

9)   Column 1: trachea
        Column 2: respiratory

        Answer: respiratory
        Type: MA   Page Ref: 7
        Topic: SYSTEMS: Organs and Functions

10)   Column 1: esophagus
        Column 2: digestive

        Answer: digestive
        Type: MA   Page Ref: 8
        Topic: SYSTEMS: Organs and Functions

11)   Column 1: salivary glands
        Column 2: digestive

        Answer: digestive
        Type: MA   Page Ref: 8
        Topic: SYSTEMS: Organs and Functions

12)   Column 1: ureter
        Column 2: urinary

        Answer: urinary
        Type: MA   Page Ref: 8
        Topic: SYSTEMS: Organs and Functions

13) Column 1: female urethra
Column 2: urinary

Answer: urinary
Type: MA   Page Ref: 8
Topic: SYSTEMS: Organs and Functions

14) Column 1: ovaries
Column 2: reproductive

Answer: reproductive
Type: MA   Page Ref: 9
Topic: SYSTEMS: Organs and Functions

15) Column 1: testes
Column 2: reproductive

Answer: reproductive
Type: MA   Page Ref: 9
Topic: SYSTEMS: Organs and Functions

16) Column 1: heart
Column 2: cardiovascular

Answer: cardiovascular
Type: MA   Page Ref: 5
Topic: SYSTEMS: Organs and Functions

17) Column 1: larynx
Column 2: respiratory

Answer: respiratory
Type: MA   Page Ref: 7
Topic: SYSTEMS: Organs and Functions

Match the anatomical regions with their common names.

18) Column 1: cephalic
Column 2: head

Answer: head
Type: MA   Page Ref: 10
Topic: REGIONAL NAMES

19) Column 1: sternal
Column 2: breastbone

Answer: breastbone
Type: MA   Page Ref: 10
Topic: REGIONAL NAMES

20) Column 1: olecranal
Column 2: posterior surface at elbow

Answer: posterior surface at elbow
Type: MA   Page Ref: 10
Topic: REGIONAL NAMES

21) Column 1: vertebral
Column 2: spinal column

Answer: spinal column
Type: MA   Page Ref: 10
Topic: REGIONAL NAMES

22) Column 1: carpal
Column 2: wrist

Answer: wrist
Type: MA   Page Ref: 10
Topic: REGIONAL NAMES

23) Column 1: axillary
Column 2: armpit

Answer: armpit
Type: MA   Page Ref: 10
Topic: REGIONAL NAMES

24) Column 1: digital
Column 2: toes or fingers

Answer: toes or fingers
Type: MA   Page Ref: 10
Topic: REGIONAL NAMES

25) Column 1: patellar
Column 2: anterior surface at the knee

Answer: anterior surface at the knee
Type: MA   Page Ref: 10
Topic: REGIONAL NAMES

26) Column 1: tarsal
Column 2: ankle

Answer: ankle
Type: MA   Page Ref: 10
Topic: REGIONAL NAMES

27) Column 1: plantar
Column 2: sole

Answer: sole
Type: MA   Page Ref: 10
Topic: REGIONAL NAMES

28) Column 1: thoracic
Column 2: chest

Answer: chest
Type: MA   Page Ref: 10
Topic: REGIONAL NAMES

29) Column 1: coxal
Column 2: hip

Answer: hip
Type: MA   Page Ref: 10
Topic: REGIONAL NAMES

CHAPTER 2   Cells

MULTIPLE CHOICE.   Choose the one alternative that best completes the statement or answers the question.

1)   The plasma membrane contains:
       1. sugars
       2. microvilli
       3. glycolipids
       4. phospholipids
       5. cholesterol
       6. glycoproteins
       A) 1, 2, 3, 4, 5, 6   B) 2, 4, 6   C) 3, 4, 5, 6   D) 1, 3, 4, 5
       Answer: C
       Type: MC   Page Ref: 31
       Topic: PLASMA MEMBRANE: Chemistry and Anatomy

2)   Which of the following is *not* considered to be true of peripheral proteins of the plasma membrane?
       A) Some serve as cell identity markers.
       B) Some act as anchors for the cytoskeleton.
       C) Some form diffusion pores in the plasma membranes.
       D) Some act as enzymes.

       Answer: C
       Type: MC   Page Ref: 31
       Topic: PLASMA MEMBRANE: Chemistry and Anatomy

3)   Which of the following is true of the lipids of the cell membrane?
       A) Cholesterol strengthens the membrane.
       B) Glycolipids are important for cell-to-cell recognition, adhesion, and communication.
       C) Phospholipids are the most abundant lipid in the membrane, forming the bilayer or basic framework of the membrane.
       D) All of the above statements are true.

       Answer: D
       Type: MC   Page Ref: 31
       Topic: PLASMA MEMBRANE: Chemistry and Anatomy

4)   In passive processes, substances move across the plasma membrane:
       1. with the use of ATP.
       2. up their own concentration gradient.
       3. down a pressure gradient.
       4. equally in all directions.
       A) 3 only   B) 2 only   C) 1, 2, 3   D) 2, 3, 4
       Answer: A
       Type: MC   Page Ref: 31
       Topic: PLASMA MEMBRANE: Functions

5) Which of the following is/are found in the cytoplasm of a cell?
A) organelles, inclusions, cytosol
B) organelles, inclusions, nucleus
C) cytosol only
D) organelles, inclusions, cytosol, nucleus

Answer: A
Type: MC   Page Ref: 33
Topic: CYTOPLASM

6) Which of the following structures is *not* bounded by a membrane?
A) ribosome
B) lysosome
C) nucleus
D) Golgi complex

Answer: A
Type: MC   Page Ref: 35
Topic: ORGANELLES

7) Rough endoplasmic reticulum differs from smooth endoplasmic reticulum in that it has _____ associated with it, and therefore rough endoplasmic reticulum assists in production and temporary storage of _____.
A) inclusions, cytosol
B) lysosomes, lipids
C) ribosomes, lysosomes
D) ribosomes, proteins

Answer: D
Type: BI   Page Ref: 37
Topic: ORGANELLES

8) The membrane of endoplasmic reticulum is continuous with the membrane of the _____.
A) cell
B) nucleus
C) mitochondria
D) ribosomes

Answer: B
Type: BI   Page Ref: 37
Topic: ORGANELLES

9) Organelles that are similar to small lysosomes, that contain enzymes, and that are important in detoxification are called _____.
A) ribosomes
B) mitochondria
C) peroxisomes
D) centrosomes

Answer: C
Type: BI   Page Ref: 41
Topic: ORGANELLES

10) The cytoskeleton consists of:
A) microfilaments.
B) intermediate filaments.
C) microtubules.
D) all of the above.

Answer: D
Type: MC   Page Ref: 44
Topic: ORGANELLES

11) Which of the following is *not* located in the nucleus?
A) histones
B) chromatin
C) endosome
D) nucleosome

Answer: C
Type: MC   Page Ref: 33
Topic: ORGANELLES

12) Melanin, triglycerides, and glycogen are examples of substances found in cells in structures known as _____.
A) lysosomes
B) endosomes
C) pinocytic vesicles
D) inclusions

Answer: D
Type: BI   Page Ref: 44
Topic: CELL INCLUSIONS

13) A cell divides to produce two new cells. The two processes that must occur during this event are:
A) mitosis and meiosis.
B) nuclear division and cytokinesis.
C) anabolism and catabolism.
D) autophagy and endocytosis.

Answer: B
Type: MC   Page Ref: 45
Topic: SOMATIC CELL DIVISION

14) The stages of the cell cycle occur in the following order:
A) G1, S, G2, mitosis.
B) S, G1, G2, interphase.
C) G1, G2, S, mitosis.
D) G1, G2, S, interphase.

Answer: A
Type: MC   Page Ref: 47
Topic: SOMATIC CELL DIVISION

15) Which of the following events does *not* occur during prophase of mitosis?
A) The nucleolus and the nuclear membrane disintegrate.
B) Each chromosome replicates so that it becomes a double–stranded structure.
C) Centrosomes start to form the mitotic spindle.
D) Chromosomes become visible due to the shortening and condensation of chromosomes.

Answer: B
Type: MC   Page Ref: 47
Topic: MITOSIS

16) An abnormal, uncontrolled increase in the rate of mitosis in a tissue is: _____.
A) hyperplasia   B) metastasis   C) metaplasia   D) oncology

Answer: A
Type: BI   Page Ref: 51
Topic: KEY MEDICAL TERMS

17) Which of the following is considered a factor in aging?
A) increased synthesis of protein
B) viral invasion of cells
C) free radical damage to protein and DNA
D) decreased autoimmune response

Answer: C
Type: MC   Page Ref: 50
Topic: AGING

14

TRUE/FALSE. Write 'T' if the statement is true and 'F' if the statement is false.

1) Protein molecules form the basic framework of the plasma membrane.

    Answer: FALSE
    Type: TF   Page Ref: 31
    Topic: PLASMA MEMBRANE: Chemistry and Anatomy

2) Plasma and lymph are two kinds of interstitial fluid.

    Answer: FALSE
    Type: TF   Page Ref: 31
    Topic: BODY FLUIDS

3) Cytosol is the same as intracellular fluid.

    Answer: TRUE
    Type: TF   Page Ref: 31
    Topic: CYTOPLASM

4) Active transport of molecules across cell membranes requires the use of adenosine triphosphate (ATP) as energy.

    Answer: TRUE
    Type: TF   Page Ref: 32
    Topic: PLASMA MEMBRANE: Functions

5) Hormones, cholesterol, vitamins, and some viruses enter cells via the process of phagocytosis.

    Answer: FALSE
    Type: TF   Page Ref: 33
    Topic: PLASMA MEMBRANE: Functions

6) Endocytosis is the reverse of exocytosis and pinocytosis is the reverse of phagocytosis.

    Answer: FALSE
    Type: TF   Page Ref: 33
    Topic: PLASMA MEMBRANE: Functions

7) Diffusion and osmosis are passive processes that occur primarily due to the kinetic energy of molecules.

    Answer: TRUE
    Type: TF   Page Ref: 32
    Topic: PLASMA MEMBRANE: Functions

8) All cells of the body have at least one nucleus.

    Answer: FALSE
    Type: TF   Page Ref: 35
    Topic: ORGANELLES

9) Proteins produced in a cell, but destined for export from the cell, will be packaged in vesicles of the Golgi complex.

    Answer: TRUE
    Type: TF   Page Ref: 38
    Topic: ORGANELLES

10) Lysosomes contain enzymes that digest the contents of phagocytic vesicles, pinocytic vesicles, and endosomes.

Answer: TRUE
Type: TF   Page Ref: 39
Topic: ORGANELLES

11) Microtubules of the cytoskeleton are formed from the protein actin.

Answer: FALSE
Type: TF   Page Ref: 44
Topic: ORGANELLES

12) The number and kind of inclusions in a cell do not vary appreciably over the life of the cell.

Answer: FALSE
Type: TF   Page Ref: 44
Topic: CELL INCLUSIONS

13) The kinetochore is part of the mitotic spindle.

Answer: FALSE
Type: TF   Page Ref: 47
Topic: MITOSIS

14) The mitotic spindle consists of microtubules.

Answer: TRUE
Type: TF   Page Ref: 49
Topic: MITOSIS

15) A neoplasm is an example of cell division gone wild.

Answer: TRUE
Type: TF   Page Ref: 51
Topic: MITOSIS

ESSAY.   Write your answer in the space provided or on a separate sheet of paper.

1) List four functions of the cell membrane and relate each to your knowledge of cell membrane structure and chemistry.

Answer: The answer should contain the four functions found on p. 31. Relate the functions to structure as follows:

Communication: Discuss the role of proteins as receptors.

Shape and protection: The fluid mosaic model describes the flexible boundary that encloses cell contents.

Electrochemical gradient: The membrane forms a barrier that maintains an unequal arrangement of charged particles inside and outside the cell.

Selective permeability: Passage of molecules depends upon their ability to fit through pores or to dissolve in the lipids of the membrane, or upon specific carrier mechanisms.

Type: ES   Page Ref: 31
Topic: PLASMA MEMBRANE: Functions

2) Integral proteins play important roles in certain transport mechanisms. What roles do they play in osmosis, facilitated diffusion, and active transport?

Answer: Osmosis: Integral proteins form channels through which water molecules travel.
Facilitated diffusion: Integral proteins function as transporters by altering the solubility and shape of the molecules being transported.
Active transport: Integral proteins form channels and, in addition, provide a site for interaction with ATP, the source of energy for the process.
Type: ES   Page Ref: 31, 32
Topic: PLASMA MEMBRANE: Functions

3) List and define: A) four passive processes and B) three active processes that accomplish the movement of substances across cell membranes. Give an example of a molecule that is moved by each process.

Answer: Refer to Table 2.1
Type: ES   Page Ref: 32, 33
Topic: PLASMA MEMBRANE: Functions

4) For each stage of mitosis, give a two- or three-sentence description of events. Your answer should include the names of the four stages and proper use of the following terms: chromatin, chromosome, chromatid, centromere, sister chromatid, daughter chromosome.

Answer: See Table 2.3 for brief descriptions of each stage. Answer should exhibit clear understanding of the listed terms, as used in the text on pp. 47–49.
Type: ES   Page Ref: 47–50
Topic: MITOSIS

SHORT ANSWER. Write the word or phrase that best completes each statement or answers the question.

1) Microfilaments of the cytoskeleton are often anchored in place through attachment to _____ of the plasma membrane.

Answer: peripheral proteins
Type: SA   Page Ref: 31
Topic: PLASMA MEMBRANE: Chemistry and Anatomy

2) The plasma membrane may be described as two layers of _____ molecules with _____ molecules floating in their midst.

Answer: phospholipid, protein
Type: SA   Page Ref: 31
Topic: PLASMA MEMBRANE: Chemistry and Anatomy

3) In the process of _____, transport of glucose often involves integral protein transporters.

Answer: facilitated diffusion
Type: SA   Page Ref: 32
Topic: PLASMA MEMBRANE: Functions

4) Water travels continuously between interstitial fluid and intracellular fluid via the process of _____.

Answer: osmosis
Type: SA   Page Ref: 32
Topic: BODY FLUIDS

5) Two organelles, the nucleus and the _____, are enclosed in double membranes.

Answer: mitochondria
Type: SA   Page Ref: 42
Topic: ORGANELLES

6) Smooth endoplasmic reticulum is the site of production of _____.

Answer: fatty acids, phospholipids, or steroids
Type: SA   Page Ref: 37
Topic: ORGANELLES

7) Cisterns (reservoirs or sacs) is a term used to describe portions of two organelles, the _____ and the _____.

Answer: endoplasmic reticulum, Golgi complex
Type: SA   Page Ref: 37, 38
Topic: ORGANELLES

8) The process in which lysosomes digest their host cell is _____.

Answer: autolysis
Type: SA   Page Ref: 41
Topic: ORGANELLES

9) The contents of the innermost region of a mitochondrion are called the _____.

Answer: matrix
Type: SA   Page Ref: 42
Topic: ORGANELLES

10) The only human cell that has a flagellum is the _____ cell.

Answer: sperm
Type: SA   Page Ref: 44
Topic: ORGANELLES

11) _____ microtubules shorten and pull chromosomes during anaphase.

Answer: kinetochore
Type: SA   Page Ref: 49
Topic: MITOSIS

12) Somatic cell division involves the process of cytokinesis and _____.

Answer: mitosis
Type: SA   Page Ref: 45
Topic: MITOSIS

MATCHING. Choose the item in Column 2 that best matches each item in Column 1.

Match the following fluids with their specific locations.

1)   Column 1: plasma
        Column 2: fluid portion of blood

        Answer: fluid portion of blood
        Type: MA   Page Ref: 31
        Topic: BODY FLUIDS

2)   Column 1: interstitial fluid
        Column 2: found in spaces between
                  cells

        Answer: found in spaces between cells
        Type: MA   Page Ref: 31
        Topic: BODY FLUIDS

3)   Column 1: intracellular fluid
        Column 2: fluid component of
                  cytoplasm

        Answer: fluid component of cytoplasm
        Type: MA   Page Ref: 31
        Topic: BODY FLUIDS

4)   Column 1: lymph
        Column 2: carried in lymphatic vessels

        Answer: carried in lymphatic vessels
        Type: MA   Page Ref: 31
        Topic: BODY FLUIDS

5)   Column 1: extracellular fluid
        Column 2: found, for example, in
                  urinary bladder

        Answer: found, for example, in urinary bladder
        Type: MA   Page Ref: 31
        Topic: BODY FLUIDS

Match the names of the membrane transport processes with the examples.

6)   Column 1: diffusion
        Column 2: oxygen moves from air sacs
                  of lungs into blood of
                  surrounding capillaries

        Answer: oxygen moves from air sacs of lungs into blood of surrounding capillaries
        Type: MA   Page Ref: 32
        Topic: PLASMA MEMBRANE: Functions

7)   Column 1: diffusion
      Column 2: carbon dioxide moves from
               cells into the blood of
               nearby capillaries

      Answer: carbon dioxide moves from cells into the blood of nearby capillaries
      Type: MA   Page Ref: 32
      Topic: PLASMA MEMBRANE: Functions

8)   Column 1: exocytosis
      Column 2: secretion of digestive
               enzymes
      Foil: fluids move from the mouth
               to the stomach

      Answer: secretion of digestive enzymes
      Type: MA   Page Ref: 33
      Topic: PLASMA MEMBRANE: Functions

9)   Column 1: filtration
      Column 2: fluids move from the blood
               into nephrons of kidneys in
               the first stage of urine
               formation

      Answer: fluids move from the blood into nephrons of kidneys in the first stage of urine
               formation
      Type: MA   Page Ref: 32
      Topic: PLASMA MEMBRANE: Functions

10)   Column 1: osmosis
      Column 2: water moves from red blood
               cells into plasma

      Answer: water moves from red blood cells into plasma
      Type: MA   Page Ref: 32
      Topic: PLASMA MEMBRANE: Functions

11)   Column 1: phagocytosis
      Column 2: a mechanism that employs
               formation of pseudopods;
               moves solid particles into
               cells
      Foil: food passes from the mouth
               to the stomach

      Answer: a mechanism that employs formation of pseudopods; moves solid particles into
               cells
      Type: MA   Page Ref: 33
      Topic: PLASMA MEMBRANE: Functions

12) Column 1: pinocytosis
Column 2: a process employed by most
cells; moves small droplets
of extracellular fluid into
cells

Answer: a process employed by most cells; moves small droplets of extracellular fluid
into cells
Type: MA   Page Ref: 33
Topic: PLASMA MEMBRANE: Functions

13) Column 1: active transport
Column 2: maintains the unequal
distribution of potassium
and sodium ions across a
nerve cell membrane

Answer: maintains the unequal distribution of potassium and sodium ions across a
nerve cell membrane
Type: MA   Page Ref: 33
Topic: PLASMA MEMBRANE: Functions

Match the events in Column 2 with the phases of somatic cell division in Column 1.
14) Column 1: interphase
Column 2: replication of DNA and
organelles

Answer: replication of DNA and organelles
Type: MA   Page Ref: 47
Topic: MITOSIS

15) Column 1: prophase
Column 2: chromosomes are visible as
46 double-stranded
structures
Foil: mitochondrial membranes
disintegrate

Answer: chromosomes are visible as 46 double-stranded structures
Type: MA   Page Ref: 47
Topic: MITOSIS

16) Column 1: prophase
Column 2: nucleoli disappear

Answer: nucleoli disappear
Type: MA   Page Ref: 47
Topic: MITOSIS

17)   Column 1: metaphase
       Column 2: centromeres of chromosomes
               attach to spindle in the
               equatorial plane

Answer: centromeres of chromosomes attach to spindle in the equatorial plane
Type: MA   Page Ref: 49
Topic: MITOSIS

18)   Column 1: anaphase
       Column 2: centromeres divide so that
               the sister chromatids
               separate

Answer: centromeres divide so that the sister chromatids separate
Type: MA   Page Ref: 49
Topic: MITOSIS

19)   Column 1: anaphase
       Column 2: daughter chromosomes
               begin to move to opposite
               poles of the cell

Answer: daughter chromosomes begin to move to opposite poles of the cell
Type: MA   Page Ref: 49
Topic: MITOSIS

20)   Column 1: telophase
       Column 2: nuclear membrane becomes
               visible again around two
               separate masses of
               chromatin

Answer: nuclear membrane becomes visible again around two separate masses of
        chromatin
Type: MA   Page Ref: 49
Topic: MITOSIS

21)   Column 1: telophase
       Column 2: cleavage furrow deepens to
               accomplish complete cell
               division

Answer: cleavage furrow deepens to accomplish complete cell division
Type: MA   Page Ref: 49
Topic: MITOSIS

CHAPTER 3   Tissues

MULTIPLE CHOICE.   Choose the one alternative that best completes the statement or answers the question.

1) The three primary germ layers from which all tissues arise are:
A) ectoderm, endoderm, and plasma.
B) epithelium, endothelium, and mesothelium.
C) epithelium, endoderm, and mesenchyme.
D) endoderm, mesoderm, and ectoderm.

Answer: D
Type: MC   Page Ref: 58
Topic: TISSUES: Origins

2) Which of the following cell junctions anchor or connect adjacent cells to one another?
1. hemidesmosomes
2. focal adhesions
3. adhesion belts
4. desmosomes
5. gap junctions
6. tight junctions
A) all of the above   B) 2, 4, 5, 6
C) 3, 4, 5, 6   D) 4 and 6 only

Answer: C
Type: MC   Page Ref: 60, 61
Topic: CELL JUNCTIONS

3) Dermatan sulfate, adhesion proteins, hyaluronic acid, and chondroitin sulfate can be found in:
A) ground substance of connective tissues.
B) cell membranes of epithelial cells.
C) vacuoles of adipocytes.
D) ducts of salivary glands.

Answer: A
Type: MC   Page Ref: 71
Topic: CONNECTIVE TISSUE: General Features

4) Epithelial tissues can be classified into their various categories according to:
A) cell shape and number of layers.   B) type of fibers and number of layers.
C) presence or absence of blood vessels.   D) location in the body.

Answer: A
Type: MC   Page Ref: 62
Topic: EPITHELIAL TISSUE

5) Which of the following is a false statement about epithelial tissues?
A) Cell junctions are plentiful.
B) Epithelium is a highly vascular tissue.
C) Cells are replaced continuously by mitosis.
D) Epithelial functions include: protection, lubrication, digestion, and reproduction.
Answer: B
Type: MC   Page Ref: 61
Topic: EPITHELIAL TISSUE

6) Which of the following is the thinnest epithelial tissue?
A) simple cuboidal   B) simple columnar
C) simple squamous   D) transitional
Answer: C
Type: MC   Page Ref: 62
Topic: EPITHELIAL TISSUE

7) A gland may consist of:
A) a single epithelial cell.   B) a cluster of epithelial cells.
C) neither A nor B is correct.   D) both A and B are correct.
Answer: D
Type: MC   Page Ref: 59
Topic: EPITHELIAL TISSUE: Glandular Epithelium

8) In which type of exocrine gland are portions of cells containing secretions pinched off from the main cell body?
A) holocrine glands   B) merocrine glands
C) apocrine glands   D) all of the above
Answer: C
Type: MC   Page Ref: 70
Topic: EPITHELIAL TISSUE: Glandular Epithelium

9) Which is the most common type of exocrine gland?
A) holocrine glands   B) merocrine glands
C) apocrine glands   D) all of the above
Answer: B
Type: MC   Page Ref: 70
Topic: EPITHELIAL TISSUE: Glandular Epithelium

10) Which of the following is a vascularized tissue?
A) adipose tissue
B) hyaline cartilage
C) keratinized stratified squamous epithelium
D) simple cuboidal epithelium
Answer: A
Type: MC   Page Ref: 70
Topic: CONNECTIVE TISSUE: Adipose

11) Connective tissue may contain:
A) fibroblasts.
B) plasma cells.
C) macrophages and mast cells.
D) all of the above.

Answer: D
Type: MC   Page Ref: 71
Topic: CONNECTIVE TISSUE: General Features

12) The matrix of connective tissue is produced by:
A) plasma cells.
B) macrophages.
C) mast cells.
D) none of the above.

Answer: D
Type: MC   Page Ref: 71
Topic: CONNECTIVE TISSUE: General Features

13) All three types of fibers (collagen, elastic, and reticular) are found in which of the following tissues?
A) reticular connective tissue
B) hyaline cartilage
C) tendons
D) areolar connective tissue

Answer: D
Type: MC   Page Ref: 72
Topic: CONNECTIVE TISSUE: Loose

14) The flexible connective tissue that joins the anterior extremity of a rib to the breast bone or sternum is:
A) dense regular connective tissue.
B) hyaline cartilage.
C) elastic cartilage.
D) fibrocartilage.

Answer: B
Type: MC   Page Ref: 77
Topic: CONNECTIVE TISSUE: Cartilage

15) Which of the following is *not* a characteristic of bone?
A) lacunae   B) osteocytes   C) perichondrium   D) lamellae

Answer: C
Type: MC   Page Ref: 80
Topic: CONNECTIVE TISSUE: Bone

16) Which of the following are characteristics of fibrocartilage?
1. chondrocytes in lacunae
2. bundles of collagen fibers
3. perichondrium
4. canaliculi
5. lamellae
A) 1, 2, 3, 4   B) 1, 2, 3   C) 3, 4, 5   D) 1 and 2 only

Answer: D
Type: MC   Page Ref: 78
Topic: CONNECTIVE TISSUE: Cartilage

17) The tissue that contains erythrocytes, leukocytes, and plasma belongs to which tissue group?
A) epithelial tissue
B) connective tissue
C) muscle tissue
D) nervous tissue

Answer: B
Type: MC   Page Ref: 81
Topic: CONNECTIVE TISSUE: Blood

18) The central canal of an osteon contains:
A) red marrow.
B) yellow marrow.
C) blood vessels and nerves.
D) matrix.

Answer: C
Type: MC   Page Ref: 81
Topic: CONNECTIVE TISSUE: Bone

19) An epithelial membrane consists of:
A) epithelium only.
B) epithelium plus connective tissue.
C) epithelium plus blood vessels.
D) epithelium plus nervous tissue.

Answer: B
Type: MC   Page Ref: 81
Topic: MEMBRANES

20) The membrane covering the outer surface of the heart is a _____ membrane.
A) mucous   B) serous   C) synovial   D) parietal

Answer: B
Type: BI   Page Ref: 81
Topic: MEMBRANES

21) The membrane covering the outer surface of the lungs is called _____.
A) parietal pleura
B) parietal peritoneum
C) visceral pleura
D) visceral peritoneum

Answer: C
Type: BI   Page Ref: 81
Topic: MEMBRANES

22) The tissue layers that make up a mucous membrane are:
A) loose connective tissue and dense connective tissue.
B) epithelium and adipose tissue.
C) epithelium and smooth muscle.
D) epithelium and lamina propria.

Answer: D
Type: MC   Page Ref: 81
Topic: MEMBRANES

23) Skeletal muscle fibers (cells) have the following anatomical characteristic(s):
A) several nuclei.
B) striations.
C) cylindrical unbranched fibers.
D) all of the above.

Answer: D
Type: MC   Page Ref: 82
Topic: MUSCLE TISSUE

24) Cardiac muscle cells (fibers) have the following anatomical characteristic(s):
A) several nuclei.   B) parallel unbranched fibers.
C) striations.   D) all of the above.

Answer: C
Type: MC   Page Ref: 82
Topic: MUSCLE TISSUE

25) Which type of muscle tissue is involuntary and contains desmosomes and gap junctions in structures called intercalated discs?
A) skeletal   B) cardiac
C) smooth   D) all of the above

Answer: B
Type: BI   Page Ref: 82
Topic: MUSCLE TISSUE

26) Neuroglia are cells in nervous tissue that:
A) generate nerve impulses.
B) conduct nerve impulses.
C) nourish, support, and protect neurons.
D) all of the above.

Answer: C
Type: MC   Page Ref: 83
Topic: NERVOUS TISSUE

TRUE/FALSE. Write 'T' if the statement is true and 'F' if the statement is false.

1) Pseudostratified columnar epithelium has one layer of cells.

Answer: TRUE
Type: TF   Page Ref: 69
Topic: EPITHELIAL TISSUE

2) Stratified epithelium is an effective protective layer because cells in the basal layer reproduce quickly to replace surface cells that are shed.

Answer: TRUE
Type: TF   Page Ref: 68
Topic: EPITHELIAL TISSUE

3) Closely packed cells and fibers are characteristic of connective tissue.

Answer: FALSE
Type: TF   Page Ref: 70
Topic: CONNECTIVE TISSUE: General Features

4) Tendons and ligaments are examples of dense regular connective tissue.

Answer: TRUE
Type: TF   Page Ref: 76
Topic: CONNECTIVE TISSUE: Dense

5) Dense connective tissue is termed dense because of the thick consistency of the fluid ground substance.

Answer: FALSE
Type: TF   Page Ref: 76
Topic: CONNECTIVE TISSUE: Dense

6) A synovial membrane consists of a thin sheet of epithelial tissue.

Answer: FALSE
Type: TF   Page Ref: 81
Topic: MEMBRANES

7) The basic structural unit of compact bone is called an osteon.

Answer: TRUE
Type: TF   Page Ref: 80
Topic: CONNECTIVE TISSUE: Bone

8) Serous membranes line cavities of the body that are not open to the external environment.

Answer: TRUE
Type: TF   Page Ref: 81
Topic: MEMBRANES

9) The membrane that lines the cavities of freely movable joints secretes a lubricating fluid called serous fluid.

Answer: FALSE
Type: TF   Page Ref: 81
Topic: MEMBRANES

10) Intercalated discs bind smooth muscle cells together and also allow for the passage of muscle action potentials between cells.

Answer: FALSE
Type: TF   Page Ref: 82
Topic: MUSCLE TISSUE

11) The neuron is the structural and functional unit of the nervous system.

Answer: TRUE
Type: TF   Page Ref: 83
Topic: NERVOUS TISSUE

ESSAY. Write your answer in the space provided or on a separate sheet of paper.

1) Describe the epithelial group of tissues by listing ten general features that all types of epithelium have in common.

Answer: See General Features, pp. 61.
Type: ES   Page Ref: 61
Topic: EPITHELIAL TISSUE

2)   Name and describe the three functional types of multicellular exocrine glands, giving an example of each.

Answer: See pp. 70 for descriptions of the three types.
1. Holocrine: sebaceous glands of the skin
2. Merocrine: salivary glands
3. Apocrine: mammary glands
Type: ES   Page Ref: 70
Topic: EPITHELIAL TISSUE: Glandular Epithelium

3)   Compare and contrast the structure of the two types of bone tissue.

Answer: In general, these two are quite similar. Answer should include the details of the osteon of compact bone and a general description of the components of the trabeculae of spongy bone as on pp. 80, 81.
Type: ES   Page Ref: 80, 81
Topic: CONNECTIVE TISSUE: Bone

4)   List four features and one location for each of the three types of muscle tissue.

Answer: Describe each using the following terms: (1) striated or nonstriated, (2) voluntary or involuntary, (3) uninucleate or multinucleate, (4) cylindrical or branched or spindle shaped. See Table 3.4 for locations.
Type: ES   Page Ref: 82–84
Topic: MUSCLE TISSUE

SHORT ANSWER.   Write the word or phrase that best completes each statement or answers the question.

1)   Proteins that bridge an intercellular space to form a gap junction are called _____.

Answer: connexons
Type: SA   Page Ref: 61
Topic: CELL JUNCTIONS

2)   Epithelial tissues secrete the _____ layer of the basement membrane.

Answer: basal lamina
Type: SA   Page Ref: 61
Topic: EPITHELIAL TISSUE

3)   Endocrine glands secrete hormones into _____, from which they diffuse into the _____; exocrine gland secretions are released into _____.

Answer: extracellular fluid, bloodstream, ducts
Type: SA   Page Ref: 69
Topic: EPITHELIAL TISSUE: Glandular Epithelium

4)   The type of exocrine gland that releases its secretions packaged inside whole cells is called a/an _____ gland.

Answer: holocrine
Type: SA   Page Ref: 70
Topic: EPITHELIAL TISSUE: Glandular Epithelium

5) The two forms of simple columnar epithelium are _____ and _____.

Answer: ciliated, nonciliated
Type: SA   Page Ref: 64
Topic: EPITHELIAL TISSUE

6) The matrix of _____ tissue is rich in calcium and phosphorus.

Answer: bone
Type: SA   Page Ref: 80
Topic: CONNECTIVE TISSUE: Bone

7) Osteocytes reside in _____ in bone tissue.

Answer: lacunae
Type: SA   Page Ref: 80
Topic: CONNECTIVE TISSUE: Bone

8) A cell called a/an _____ consists of axon, dendrites, and a cell body.

Answer: neuron
Type: SA   Page Ref: 83
Topic: NERVOUS TISSUE

9) The three types of cartilage in the body are _____, _____, and _____.

Answer: hyaline cartilage, fibrocartilage, elastic cartilage
Type: SA   Page Ref: 78
Topic: CONNECTIVE TISSUE: Cartilage

10) Fibers made of the protein _____ are abundant in ligaments and tendons.

Answer: collagen
Type: SA   Page Ref: 76
Topic: CONNECTIVE TISSUE: Dense

11) A/an _____ membrane, consisting of a thin layer of _____ tissue, lines the cavity of a freely movable joint.

Answer: synovial, areolar connective
Type: SA   Page Ref: 81
Topic: MEMBRANES

12) Peripherally located nuclei are characteristic of _____ muscle cells.

Answer: skeletal
Type: SA   Page Ref: 82
Topic: MUSCLE TISSUE

13) The cells of _____ muscle are cylindrical, with each end of the cell tapered to a point.

Answer: smooth
Type: SA   Page Ref: 83
Topic: MUSCLE TISSUE

MATCHING. Choose the item in Column 2 that best matches each item in Column 1.

Match the epithelial tissues in Column 1 with the descriptions in Column 2.

1)   Column 1: simple squamous epithelium
     Column 2: called endothelium in
               certain locations

     Answer: called endothelium in certain locations
     Type: MA   Page Ref: 62
     Topic: EPITHELIAL TISSUE

2)   Column 1: simple cuboidal epithelium
     Column 2: functions in secretion and
               absorption

     Answer: functions in secretion and absorption
     Type: MA   Page Ref: 62
     Topic: EPITHELIAL TISSUE

3)   Column 1: pseudostratified columnar
               epithelium
     Column 2: location of nuclei in cells is
               varied; not all cells reach
               the apical surface of the
               tissue layer

     Answer: location of nuclei in cells is varied; not all cells reach the apical surface of the
             tissue layer
     Type: MA   Page Ref: 67
     Topic: EPITHELIAL TISSUE

4)   Column 1: transitional epithelium
     Column 2: permits distention of hollow
               structures

     Answer: permits distention of hollow structures
     Type: MA   Page Ref: 66
     Topic: EPITHELIAL TISSUE

5)   Column 1: simple columnar epithelium
     Column 2: may include goblet cells

     Answer: may include goblet cells
     Type: MA   Page Ref: 64
     Topic: EPITHELIAL TISSUE

6)   Column 1: simple cuboidal epithelium
     Column 2: spherical, centrally located
               nucleus

     Answer: spherical, centrally located nucleus
     Type: MA   Page Ref: 63
     Topic: EPITHELIAL TISSUE

7)   Column 1: simple squamous epithelium
        Column 2: suited for diffusion,
                    osmosis, filtration

        Answer: suited for diffusion, osmosis, filtration
        Type: MA   Page Ref: 63
        Topic: EPITHELIAL TISSUE

8)   Column 1: stratified squamous
                    epithelium
        Column 2: functions in protection
                    against abrasion

        Answer: functions in protection against abrasion
        Type: MA   Page Ref: 65
        Topic: EPITHELIAL TISSUE

Match the cell junctions with the characteristics.
9)   Column 1: tight junctions
        Column 2: common in the epithelial
                    lining of the stomach,
                    intestines, and urinary
                    bladder

        Answer: common in the epithelial lining of the stomach, intestines, and urinary bladder
        Type: MA   Page Ref: 60
        Topic: CELL JUNCTIONS

10)   Column 1: plaque–bearing junctions
        Column 2: fasten cells within a tissue,
                    but allow some movement
                    of cells as an organ is
                    stretched

        Answer: fasten cells within a tissue, but allow some movement of cells as an organ is
                    stretched
        Type: MA   Page Ref: 60
        Topic: CELL JUNCTIONS

11)   Column 1: gap junctions
        Column 2: allow electrical signals to
                    travel rapidly through
                    cardiac muscle

        Answer: allow electrical signals to travel rapidly through cardiac muscle
        Type: MA   Page Ref: 61
        Topic: CELL JUNCTIONS

Match each connective tissue in Column 1 with the corresponding description of matrix in Column 2.

12)   Column 1: areolar
       Column 2: fluid, semifluid, or
                 gelatinous; randomly
                 arranged elastic, collagen,
                 and reticular fibers

       Answer: fluid, semifluid, or gelatinous; randomly arranged elastic, collagen, and reticular
               fibers
       Type: MA   Page Ref: 72
       Topic: CONNECTIVE TISSUE: Loose

13)   Column 1: dense regular
       Column 2: contains densely packed,
                 parallel bundles of collagen
                 fibers

       Answer: contains densely packed, parallel bundles of collagen fibers
       Type: MA   Page Ref: 76
       Topic: CONNECTIVE TISSUE: Dense

14)   Column 1: fibrocartilage
       Column 2: clearly visible bundles of
                 collagen fibers surrounding
                 chondrocytes in lacunae

       Answer: clearly visible bundles of collagen fibers surrounding chondrocytes in lacunae
       Type: MA   Page Ref: 78
       Topic: CONNECTIVE TISSUE: Cartilage

15)   Column 1: compact bone
       Column 2: each layer or lamella
                 contains collagen fibers that
                 provide strength and
                 mineral salts that provide
                 hardness

       Answer: each layer or lamella contains collagen fibers that provide strength and mineral
               salts that provide hardness
       Type: MA   Page Ref: 80
       Topic: CONNECTIVE TISSUE: Bone

16)   Column 1: blood
       Column 2: watery matrix called plasma

       Answer: watery matrix called plasma
       Type: MA   Page Ref: 81
       Topic: CONNECTIVE TISSUE: Blood

CHAPTER 4   The Integumentary System

MULTIPLE CHOICE. Choose the one alternative that best completes the statement or answers the question.

1)   The skin, hair, nails, glands, and cutaneous sense organs are together an example of a/an:
      A) tissue.   B) organ.   C) system.   D) organism.

      Answer: C
      Type: MC   Page Ref: 91
      Topic: SKIN: Anatomy

2)   Stratum basale contains cells named:
      A) cuboidal or columnar epithelium.   B) keratinocytes.
      C) melanocytes.   D) all of the above.

      Answer: D
      Type: MC   Page Ref: 91
      Topic: SKIN: Anatomy

3)   Cells in the lowest layer of the epidermis divide to produce all other epidermal cells, which then move progressively outward through the following layers:
      1. stratum spinosum
      2. stratum basale
      3. stratum corneum
      4. stratum lucidum (only in palms and soles)
      5. stratum granulosum
      A) 1, 2, 3, 4, 5   B) 2, 3, 1, 5, 4   C) 3, 2, 4, 1, 5   D) 2, 1, 5, 4, 3

      Answer: D
      Type: MC   Page Ref: 94
      Topic: SKIN: Anatomy

4)   As keratinocytes move to the surface:
      A) they become multinucleate.
      B) they divide to produce new skin cells.
      C) they eventually die and are sloughed off.
      D) all of the above.

      Answer: C
      Type: MC   Page Ref: 94
      Topic: SKIN: Anatomy

5)   Which of the following statements about the dermis are true?
      1. There are two layers, the upper papillary layer and the lower reticular layer.
      2. Cells found in the dermis include fibroblasts, macrophages, and keratinocytes.
      3. Dermal papillae which project upward into the epidermis commonly contain capillaries and corpuscles of touch (Meissner's corpuscles).
      A) 1 and 2   B) 1 and 3   C) 2 and 3   D) 1, 2, and 3

      Answer: B
      Type: MC   Page Ref: 95
      Topic: SKIN: Anatomy

6) The ability of skin to stretch and recoil is due to the presence of _____ in the
   _____.

   A) fibers, epidermis   B) adipose tissue, dermis
   C) fibers, dermis   D) basement membrane, hypodermis

   Answer: C
   Type: MC   Page Ref: 95
   Topic: SKIN: Anatomy

7) Functions of the skin include:
   A) excretion of some wastes.   B) blood reservoir.
   C) immunity.   D) all of the above.

   Answer: D
   Type: MC   Page Ref: 95, 96
   Topic: SKIN: Functions

8) Which of the following is *not* considered an epidermal derivative?
   A) sebaceous glands   B) lamellated corpuscles
   C) sudoriferous glands   D) hair

   Answer: B
   Type: MC   Page Ref: 95
   Topic: EPIDERMAL DERIVATIVES

9) The narrow strip of epidermis found on the surface of the proximal border of a nail
   is called _____.

   A) eponychium   B) matrix   C) lunula   D) papilla

   Answer: A
   Type: BI   Page Ref: 100
   Topic: EPIDERMAL DERIVATIVES

10) The mesoderm of the embryo gives rise to:
    A) glands.   B) dermis.   C) hair.   D) nails.

    Answer: B
    Type: MC   Page Ref: 101
    Topic: DEVELOPMENTAL ANATOMY

TRUE/FALSE. Write 'T' if the statement is true and 'F' if the statement is false.

1) The two principal layers of skin are the epidermis and the subcutaneous layer.
   Answer: FALSE
   Type: TF   Page Ref: 91
   Topic: SKIN: Anatomy

2) One of the factors responsible for variation in skin color in different regions of the
   body is the number of melanocytes in the epidermis.
   Answer: TRUE
   Type: TF   Page Ref: 96
   Topic: SKIN COLOR

3)   Albinism is due to a total lack of melanocytes.
        Answer: FALSE
        Type: TF   Page Ref: 96
        Topic: SKIN COLOR

4)   The skin is supplied with blood by two arterial plexuses, the papillary plexus and
        the cutaneous plexus.
        Answer: TRUE
        Type: TF   Page Ref: 100
        Topic: SKIN: Anatomy

5)   The deepest layer of the skin is called the hypodermis.
        Answer: FALSE
        Type: TF   Page Ref: 91
        Topic: SKIN: Anatomy

6)   A callus is a thickening of the dermis caused by excessive friction.
        Answer: FALSE
        Type: TF   Page Ref: 91
        Topic: SKIN: Anatomy

7)   The cortex forms the major part of the shaft of the hair and it helps determine the
        color of hair.
        Answer: TRUE
        Type: TF   Page Ref: 97
        Topic: EPIDERMAL DERIVATIVES

8)   Inflamed sebaceous glands are the cause of acne.
        Answer: TRUE
        Type: TF   Page Ref: 104
        Topic: EPIDERMAL DERIVATIVES

9)   Fingernails consist of hardened glandular secretion from the tissue of the nail root.
        Answer: FALSE
        Type: TF   Page Ref: 100
        Topic: EPIDERMAL DERIVATIVES

10)   Vitamin D synthesis begins in the skin, as a result of UV radiation altering a
        precursor molecule.
        Answer: TRUE
        Type: TF   Page Ref: 96
        Topic: SKIN: Functions

11)   Vernix caseosa is secreted by sebaceous glands to protect the fetus from amniotic
        fluid.
        Answer: TRUE
        Type: TF   Page Ref: 102
        Topic: AGING AND THE INTEGUMENTARY SYSTEM

ESSAY. Write your answer in the space provided or on a separate sheet of paper.

1) Describe the tissues and structures of the two regions (layers) of the dermis.

Answer: Papillary layer: upper 1/5; areolar tissue with elastic fibers, dermal papillae, blood vessels, corpuscles of touch.
Reticular layer: lower 4/5; network of dense irregular connective tissue, elastic and collagen fibers, adipose tissue, glands, hair follicles, blood vessels, temperature receptors.
Type: ES   Page Ref: 95
Topic: SKIN: Anatomy

2) What two structures in the skin play a role in temperature regulation? Describe how each performs its function.

Answer: 1. sudoriferous glands release sweat, which evaporates, removing heat from the skin surface.
2. dermal blood vessels may carry more or less heat to the skin depending on whether they are dilated or constricted.
Type: ES   Page Ref: 95
Topic: SKIN: Functions

3) Describe the changes that occur in skin with age.

Answer: changes in characteristics of elastic and collagen fibers; decreased melanocytes, fibroblasts, and Langerhans cells; diminished function of glands; loss of subcutaneous fat; thinning of dermis; decreased rate of replacement of epidermal cells, etc., as described on pp. 102.
Type: ES   Page Ref: 102
Topic: AGING AND THE INTEGUMENTARY SYSTEM

SHORT ANSWER. Write the word or phrase that best completes each statement or answers the question.

1) The layer of epidermis with the most layers of cells is the stratum _____.

Answer: corneum
Type: SA   Page Ref: 94
Topic: SKIN: Anatomy

2) Surface patterns that develop on the skin of the hands and feet are due to the presence of dermal papillae and are called _____ ridges.

Answer: epidermal
Type: SA   Page Ref: 97
Topic: SKIN: Anatomy

3) Lipids produced by _____ granules of keratinocytes are responsible for waterproofing the skin.

Answer: lamellar
Type: SA   Page Ref: 94
Topic: SKIN: Anatomy

4) Keratohyalin granules are characteristic of the stratum _____ layer of the epidermis.

Answer: granulosum
Type: SA   Page Ref: 94
Topic: SKIN: Anatomy

5) Three pigments that contribute to the color of skin are _____, _____, and _____.

Answer: melanin, carotene, hemoglobin
Type: SA   Page Ref: 96
Topic: SKIN COLOR

6) "Goose bumps" on the skin are due to contraction of smooth muscles called _____, which are associated with hair follicles.

Answer: arrectores pilorum
Type: SA   Page Ref: 99
Topic: EPIDERMAL DERIVATIVES

7) The _____ of the hair is the portion that projects from the surface of the skin.

Answer: shaft
Type: SA   Page Ref: 97
Topic: EPIDERMAL DERIVATIVES

8) Chemical sex attractants found in secretions of sebaceous and sudoriferous glands are called _____.

Answer: pheromones
Type: SA   Page Ref: 97
Topic: EPIDERMAL DERIVATIVES

9) The layers of a hair from external to internal are _____, _____, and _____.

Answer: cuticle, cortex, medulla
Type: SA   Page Ref: 97
Topic: EPIDERMAL DERIVATIVES

10) Epithelial cells from the _____ give rise to the sebaceous glands during fetal development.

Answer: hair follicles
Type: SA   Page Ref: 101
Topic: DEVELOPMENTAL ANATOMY

11) In the fifth or sixth month, hair follicles produce delicate fetal hair called _____, which is usually shed prior to birth.

Answer: lanugo
Type: SA   Page Ref: 101
Topic: DEVELOPMENTAL ANATOMY

MATCHING.   Choose the item in Column 2 that best matches each item in Column 1.

Match the epidermal cells in Column 1 with the characteristics and functions in Column 2.

1)   Column 1: melanocytes
     Column 2: have cell processes that
                transfer pigment granules to
                keratinocytes

     Answer: have cell processes that transfer pigment granules to keratinocytes
     Type: MA   Page Ref: 91
     Topic: SKIN: Anatomy

2)   Column 1: Langerhans cells
     Column 2: produced in bone marrow;
                migrate to the epidermis,
                where they interact with
                helper T cells in the
                immune response

     Answer: produced in bone marrow; migrate to the epidermis, where they interact with
             helper T cells in the immune response
     Type: MA   Page Ref: 91
     Topic: SKIN: Anatomy

3)   Column 1: Merkel cells
     Column 2: respond to mechanical
                stimulation in hairless skin

     Answer: respond to mechanical stimulation in hairless skin
     Type: MA   Page Ref: 91
     Topic: SKIN: Anatomy

4)   Column 1: keratinocytes
     Column 2: produce protein keratin,
                that provides protection
                from light, heat, microbes,
                and many chemicals

     Answer: produce protein keratin, that provides protection from light, heat, microbes, and
             many chemicals
     Type: MA   Page Ref: 91
     Topic: SKIN: Anatomy

Match the names of epithelial layers of the epidermis in Column 1 with their characteristics in
Column 2.

5)   Column 1: stratum spinosum
     Column 2: 8-10 rows of many-sided
                keratinocytes held together
                by desmosomes

     Answer: 8-10 rows of many-sided keratinocytes held together by desmosomes
     Type: MA   Page Ref: 94
     Topic: SKIN: Anatomy

6)   Column 1: stratum lucidum
        Column 2: present only in thick skin
                  of palms and soles

        Answer: present only in thick skin of palms and soles
        Type: MA   Page Ref: 94
        Topic: SKIN: Anatomy

7)   Column 1: stratum basale
        Column 2: contains keratinocytes that
                  undergo mitosis

        Answer: contains keratinocytes that undergo mitosis
        Type: MA   Page Ref: 93
        Topic: SKIN: Anatomy

8)   Column 1: stratum corneum
        Column 2: several layers of flat, dead,
                  keratinized cells

        Answer: several layers of flat, dead, keratinized cells
        Type: MA   Page Ref: 94
        Topic: SKIN: Anatomy

9)   Column 1: stratum granulosum
        Column 2: 3-5 rows of flattened
                  keratinocytes; cell organelles
                  begin to degenerate and
                  intermediate filaments
                  become organized into
                  coarse bundles

        Answer: 3-5 rows of flattened keratinocytes; cell organelles begin to degenerate and
                intermediate filaments become organized into coarse bundles
        Type: MA   Page Ref: 94
        Topic: SKIN: Anatomy

Match the characteristics in Column 1 with the appropriate type of gland in Column 2.

10)   Column 1: usually open into hair
                  follicles
        Column 2: sebaceous glands

        Answer: sebaceous glands
        Type: MA   Page Ref: 99
        Topic: EPIDERMAL DERIVATIVES

11)   Column 1: secrete oily substance called
                  sebum
        Column 2: sebaceous glands

        Answer: sebaceous glands
        Type: MA   Page Ref: 99
        Topic: EPIDERMAL DERIVATIVES

12) Column 1: most abundant in palms
     and soles
    Column 2: sudoriferous glands

    Answer: sudoriferous glands
    Type: MA   Page Ref: 99
    Topic: EPIDERMAL DERIVATIVES

13) Column 1: provide a cooling
     mechanism
    Column 2: sudoriferous glands

    Answer: sudoriferous glands
    Type: MA   Page Ref: 99
    Topic: EPIDERMAL DERIVATIVES

14) Column 1: found only in the ear canal
    Column 2: ceruminous glands

    Answer: ceruminous glands
    Type: MA   Page Ref: 100
    Topic: EPIDERMAL DERIVATIVES

15) Column 1: produce a waxy secretion
    Column 2: ceruminous glands

    Answer: ceruminous glands
    Type: MA   Page Ref: 100
    Topic: EPIDERMAL DERIVATIVES

CHAPTER 5   Bone Tissue

MULTIPLE CHOICE. Choose the one alternative that best completes the statement or answers the question.

1)   Functions of the skeletal system do *not* include:
A) protection of vital organs such as heart, lungs, and brain.
B) blood cell production.
C) control of body temperature.
D) energy storage in the form of adipose tissue.
Answer: C
Type: MC   Page Ref: 108
Topic: FUNCTIONS OF BONE

2)   Which of the following are true of yellow marrow?
1. site of blood cell production
2. found in medullary cavity of long bones
3. found in hipbones, sternum, ribs, and vertebrae
4. site of energy (fat) storage
5. can convert to red marrow under certain conditions
A) 1, 3   B) 1, 2   C) 1, 2, 5   D) 2, 4, 5
Answer: D
Type: MC   Page Ref: 108
Topic: FUNCTIONS OF BONE

3)   Which of the following is *not* true of periosteum?
A) It consists of two layers, the inner osteogenic and outer fibrous layers.
B) It assists in fracture repair.
C) It covers and protects the articular cartilages.
D) It serves as a point of attachment for tendons and ligaments.
Answer: C
Type: MC   Page Ref: 108
Topic: ANATOMY: Structure of Bone

4)   Osteocytes are the mature bone cells that develop directly from _____.
A) osteoprogenitor cells   B) osteoblasts
C) osteoclasts   D) white blood cells
Answer: B
Type: BI   Page Ref: 110
Topic: HISTOLOGY OF BONE TISSUE

5)   The hardness of bone is due to the salts of calcium and phosphate. The strength of bone is due to organic molecules such as collagen.
A) Both statements are true.
B) The first statement is true; the second is false.
C) The first statement is false; the second is true.
D) Both statements are false.
Answer: A
Type: MC   Page Ref: 110
Topic: ANATOMY: Structure of Bone

6)   Which of the following is *not* a characteristic of spongy bone tissue?
     A) lamellae   B) canaliculi
     C) osteocytes in lacunae   D) perforating (Volkmann's) canals

     Answer: D
     Type: MC   Page Ref: 112
     Topic: HISTOLOGY OF BONE TISSUE

7)   Which one of the following is characteristic of spongy bone tissue, but *not* of
     compact bone tissue?
     A) trabeculae   B) yellow marrow
     C) osteocytes in lacunae   D) interstitial lamellae

     Answer: A
     Type: MC   Page Ref: 112
     Topic: HISTOLOGY OF BONE TISSUE

8)   Intramembranous ossification is the process that:
     A) produces most bones.
     B) produces *only* flat bones of the cranium.
     C) results in growth in length of long bones.
     D) none of the above.

     Answer: D
     Type: MC   Page Ref: 113
     Topic: OSSIFICATION

9)   Put the following in correct order for endochondral ossification.
     1. Mesenchymal cells of the embryo develop into cartilage-producing cells.
     2. The periosteum (formerly perichondrium) begins to produce a thin layer of compact
     bone.
     3. A hyaline cartilage model of the future bone is formed.
     4. Cartilage in the midregion of the model becomes calcified.
     5. Spongy bone tissue develops at the primary ossification center.
     6. Secondary ossification centers produce spongy bone tissue of the epiphyses.
     7. Medullary cavity is formed.
     A) 1, 3, 4, 2, 5, 7, 6   B) 7, 2, 3, 1, 4, 5, 6
     C) 3, 1, 2, 6, 7, 4, 5   D) 1, 3, 2, 4, 6, 5, 7

     Answer: A
     Type: MC   Page Ref: 113-116
     Topic: OSSIFICATION

10) Appearance of the epiphyseal line means:
     A) the end of lengthwise growth of that bone.
     B) total replacement of epiphyseal plate by bone.
     C) all chondrocytes of the epiphyseal plate are dead.
     D) all of the above.

     Answer: D
     Type: MC   Page Ref: 117
     Topic: BONE GROWTH

11)	Growth in length of a long bone is called _____ growth.
	A) intramembranous			B) interstitial
	C) appositional				D) periosteal

	Answer: B
	*Type: BI		Page Ref: 117*
	Topic: BONE GROWTH

12)	The nutrient artery of a long bone:
	A) divides into branches that supply the marrow and the inner portion of the diaphysis.
	B) travels through Volkmann's (perforating) canals.
	C) supplies the marrow and bony tissue of the epiphysis.
	D) all of the above.

	Answer: A
	*Type: MC		Page Ref: 117*
	Topic: BLOOD AND NERVE SUPPLY

13)	Normal bone growth and replacement depends on the presence of:
	A) the vitamins A, B_{12}, C, and D.
	B) the minerals calcium, phosphorus, magnesium, boron, and manganese.
	C) calcitonin, parathyroid hormone, human growth hormone, sex hormones, and thyroid hormones.
	D) all of the above.

	Answer: D
	*Type: MC		Page Ref: 117*
	Topic: BONE REPLACEMENT

14)	What is the correct progression of events in repair of a bone fracture?
	1. development of an actively growing connective tissue procallus
	2. formation of a fracture hematoma
	3. remodeling of the callus
	4. development of a fibrocartilaginous callus
	5. development of a bony callus of spongy bone tissue
	A) 4, 1, 2, 5, 3		B) 2, 4, 3, 1, 5		C) 1, 2, 4, 3, 5		D) 2, 1, 4, 5, 3

	Answer: D
	*Type: MC		Page Ref: 118*
	Topic: FRACTURE AND REPAIR OF BONE

15)	A major change in bone tissue that occurs with aging and that leads to increased brittleness is _____.
	A) demineralization			B) decreased protein synthesis
	C) increased bone remodeling		D) all of the above

	Answer: B
	*Type: BI		Page Ref: 120*
	Topic: AGING

16) In fetal development, mesenchymal (mesodermal) cells develop into:
    A) cartilage matrix.
    B) bony matrix.
    C) chondroblasts or osteoblasts.
    D) ectodermal cells.

    Answer: C
    Type: MC   Page Ref: 120
    Topic: DEVELOPMENT

17) Osteoporosis may weaken bone to the extent that a small stress will result in a fracture. Such fractures are referred to as _____ fractures.
    A) partial   B) closed   C) pathologic   D) complete

    Answer: C
    Type: BI   Page Ref: 119
    Topic: FRACTURE AND REPAIR OF BONE

TRUE/FALSE. Write 'T' if the statement is true and 'F' if the statement is false.

1) The connective tissue found on the articular surface at the end of a bone is called endosteum.

    Answer: FALSE
    Type: TF   Page Ref: 108
    Topic: ANATOMY: Structure of Bone

2) Bone matrix contains crystallized mineral salts called hydroxyapatite.

    Answer: TRUE
    Type: TF   Page Ref: 110
    Topic: ANATOMY: Structure of Bone

3) In matrix formation, calcification precedes the secretion of collagen by osteoblasts.

    Answer: FALSE
    Type: TF   Page Ref: 110
    Topic: HISTOLOGY OF BONE TISSUE

4) Microscopic canals that run longitudinally through bone tissue and that contain blood vessels and nerves are called perforating (Volkmann's) canals.

    Answer: FALSE
    Type: TF   Page Ref: 111
    Topic: HISTOLOGY OF BONE TISSUE

5) The periosteum of bone develops from the perichondrium of the cartilage model.

    Answer: TRUE
    Type: TF   Page Ref: 114
    Topic: HISTOLOGY OF BONE TISSUE

6) Most bones develop in a process whereby hyaline cartilage models are replaced by bone tissue.

    Answer: TRUE
    Type: TF   Page Ref: 113
    Topic: OSSIFICATION

7) As a long bone grows in length, new cartilage cells are produced on the epiphyseal side of the epiphyseal plate and bone replaces cartilage on the diaphyseal side of the plate.

Answer: TRUE
Type: TF   Page Ref: 116
Topic: BONE GROWTH

8) Following completion of ossification, bone replacement occurs only if bone tissue is injured.

Answer: FALSE
Type: TF   Page Ref: 117
Topic: BONE REPLACEMENT

9) A Colles' fracture involves the distal end of the lateral forearm bone (radius).

Answer: FALSE
Type: TF   Page Ref: 119
Topic: FRACTURE AND REPAIR OF BONE

10) Limb buds appear at about the tenth week of fetal development.

Answer: FALSE
Type: TF   Page Ref: 120
Topic: DEVELOPMENT

11) At birth, all bones are cartilaginous. Ossification occurs only after birth.

Answer: FALSE
Type: TF   Page Ref: 120
Topic: DEVELOPMENT

ESSAY. Write your answer in the space provided or on a separate sheet of paper.

1) The functions of bone tissue are the functions of the skeletal system. List six functions of bone tissue.

Answer: support, protection, movement, mineral storage and release, site of blood cell production, storage of energy
Type: ES   Page Ref: 108
Topic: FUNCTIONS OF BONE

2) Briefly describe the structure and function of each of the four zones of cartilage of the epiphyseal plate.

Answer: 1. Resting: small, scattered chondrocytes; anchors the plate to the epiphysis
2. Proliferating: stacks of chondrocytes divide and replace dead cells, causing growth in length
3. Hypertrophic (maturing): columns of larger chondrocytes; maturation of cells from (2) above
4. Calcified: thin layer of dead cells in calcified matrix; attaches plate to bone of diaphysis
Type: ES   Page Ref: 116
Topic: BONE GROWTH

3)   What factors are necessary for normal bone growth and replacement?

Answer: Dietary factors: certain minerals and vitamins
         Hormonal factors: e.g. sex hormones, insulin, human growth hormone, etc.
         Mechanical stress: e.g. exercise

Type: ES   Page Ref: 117
Topic: BONE REPLACEMENT

4)   Describe the steps in fracture repair.

Answer: 1. A clot called a fracture hematoma develops. Swelling and inflammation
           occur. Phagocytes and osteoclasts begin the removal of the damaged tissue.
           2. The fracture hematoma is converted into a connective tissue procallus.
           Fibroblasts and osteoprogenitor cells invade the procallus and produce collagen
           and cartilage, converting the procallus into a cartilaginous callus.
           3. Osteoblasts develop and begin producing spongy bone tissue, thus converting
           the callus into a bony callus.
           4. The process of remodeling removes remaining dead fragments and produces
           compact bone tissue at the periphery of the fracture site.

Type: ES   Page Ref: 118
Topic: FRACTURE AND REPAIR OF BONE

SHORT ANSWER.   Write the word or phrase that best completes each statement or answers the question.

1)   The medullary cavities of long bones contain _____ marrow.

Answer: yellow
Type: SA   Page Ref: 108, 109
Topic: ANATOMY: Structure of Bone

2)   The region of a long bone where the epiphysis and diaphysis join is called the _____.

Answer: metaphysis
Type: SA   Page Ref: 108
Topic: ANATOMY: Structure of Bone

3)   The spaces in bone tissue that contain osteocytes are called _____.

Answer: lacunae
Type: SA   Page Ref: 111
Topic: HISTOLOGY OF BONE TISSUE

4)   The process by which bone tissue replaces hyaline cartilage is _____ ossification.

Answer: endochondral
Type: SA   Page Ref: 113
Topic: OSSIFICATION

5)   Flat bones that form the roof of the skull and the lower jawbone are produced by the process of _____ ossification.

Answer: intramembranous
Type: SA   Page Ref: 113
Topic: OSSIFICATION

6) The lifelong replacement and redistribution of bone matrix is referred to as _____.
    Answer: remodeling or reshaping
    Type: SA   Page Ref: 117
    Topic: BONE REPLACEMENT

7) The largest artery supplying a long bone is the _____ artery.
    Answer: nutrient
    Type: SA   Page Ref: 117
    Topic: BLOOD AND NERVE SUPPLY

8) Mechanical stress on bone is due to the pull of skeletal muscles and _____.
    Answer: gravity
    Type: SA   Page Ref: 120
    Topic: EXERCISE

9) A bone fracture in which the broken ends of the bone can be seen protruding from the skin is called a/an _____ fracture.
    Answer: open/compound
    Type: SA   Page Ref: 119
    Topic: FRACTURE AND REPAIR OF BONE

MATCHING. Choose the item in Column 2 that best matches each item in Column 1.

Match the bone cells in Column 1 with their characteristics in Column 2.

1) Column 1: osteocytes
    Column 2: maintain the daily
              metabolism of bone tissue

    Answer: maintain the daily metabolism of bone tissue
    Type: MA   Page Ref: 110
    Topic: HISTOLOGY OF BONE TISSUE

2) Column 1: osteoblasts
    Column 2: responsible for the
              formation of matrix

    Answer: responsible for the formation of matrix
    Type: MA   Page Ref: 110
    Topic: HISTOLOGY OF BONE TISSUE

3) Column 1: osteoprogenitor cells
    Column 2: undergo mitosis and
              develop into osteoblasts
    Foil: undergo mitosis and
              develop into osteoclasts

    Answer: undergo mitosis and develop into osteoblasts
    Type: MA   Page Ref: 110
    Topic: HISTOLOGY OF BONE TISSUE

4)   Column 1: osteoclasts
     Column 2: responsible for the
                 destruction (resorption) of
                 matrix

     Answer: responsible for the destruction (resorption) of matrix
     Type: MA   Page Ref: 110
     Topic: HISTOLOGY OF BONE TISSUE

5)   Column 1: osteocytes
     Column 2: mature osteoblasts

     Answer: mature osteoblasts
     Type: MA   Page Ref: 113
     Topic: HISTOLOGY OF BONE TISSUE

6)   Column 1: osteoclasts
     Column 2: release calcium from bone
                 into the bloodstream

     Answer: release calcium from bone into the bloodstream
     Type: MA   Page Ref: 113
     Topic: HISTOLOGY OF BONE TISSUE

Match the features of bone in Column 1 with the corresponding descriptive phrases in Column 2.

7)   Column 1: red marrow
     Column 2: located in spaces between
                 trabeculae

     Answer: located in spaces between trabeculae
     Type: MA   Page Ref: 112
     Topic: HISTOLOGY OF BONE TISSUE

8)   Column 1: canaliculi
     Column 2: allow movement of
                 nutrients between osteocytes

     Answer: allow movement of nutrients between osteocytes
     Type: MA   Page Ref: 111
     Topic: HISTOLOGY OF BONE TISSUE

9)   Column 1: medullary cavity
     Column 2: region of bone that
                 contains yellow marrow
     Foil: region of bone that
                 contains red marrow

     Answer: region of bone that contains yellow marrow
     Type: MA   Page Ref: 108
     Topic: HISTOLOGY OF BONE TISSUE

10)   Column 1: perforating canals
Column 2: allow blood vessels and
           nerves to penetrate compact
           bone tissue

Answer: allow blood vessels and nerves to penetrate compact bone tissue
Type: MA   Page Ref: 111
Topic: HISTOLOGY OF BONE TISSUE

11)   Column 1: osteon
Column 2: structural unit of compact
           bone

Answer: structural unit of compact bone
Type: MA   Page Ref: 111
Topic: HISTOLOGY OF BONE TISSUE

12)   Column 1: lamella
Column 2: a ring of matrix
Foil: thin plate of bone in
           spongy bone

Answer: a ring of matrix
Type: MA   Page Ref: 111
Topic: HISTOLOGY OF BONE TISSUE

13)   Column 1: metaphysis
Column 2: in a young bone, this
           region is the site of growth
           in length

Answer: in a young bone, this region is the site of growth in length
Type: MA   Page Ref: 108
Topic: HISTOLOGY OF BONE TISSUE

14)   Column 1: periosteum
Column 2: essential for growth in
           diameter of a long bone

Answer: essential for growth in diameter of a long bone
Type: MA   Page Ref: 108
Topic: HISTOLOGY OF BONE TISSUE

CHAPTER 6   The Skeletal System: The Axial Skeleton

MULTIPLE CHOICE. Choose the one alternative that best completes the statement or answers the question.

1) Spongy bone is located:
    A) in the epiphyses of long bones.   B) in the diploe of flat bones.
    C) in the ribs and sternum.   D) all of the above.

Answer: D
Type: MC   Page Ref: 126
Topic: TYPES OF BONES

2) Bones may be classified into four categories according to their shape. The four categories are:
    A) long, short, irregular, and sutural.   B) flat, irregular, long, and short.
    C) sutural, sesamoid, short, and flat.   D) short, irregular, sesamoid, and long.

Answer: B
Type: MC   Page Ref: 126
Topic: TYPES OF BONES

3) The girdles attach the limbs (extremities) to the axial skeleton. The girdles are considered to be part of the axial skeleton.
    A) Both statements are true.
    B) Both statements are false.
    C) The first statement is true; the second is false.
    D) The second statement is true; the first is false.

Answer: C
Type: MC   Page Ref: 126
Topic: DIVISIONS OF THE SKELETAL SYSTEM

4) Membrane–filled spaces between cranial bones of an infant skull are called:
    A) sutures.   B) sinuses.   C) fontanels.   D) foramina.

Answer: C
Type: MC   Page Ref: 148
Topic: SKULL

5) The frontal bone:
    A) contributes to the sagittal suture.   B) forms part of the floor of the orbit.
    C) articulates with the mandible.   D) none of the above.

Answer: D
Type: MC   Page Ref: 131
Topic: SKULL

6) Which skull bone contains the mandibular fossa, mastoid process, styloid process, and zygomatic process?
    A) temporal   B) parietal   C) zygomatic   D) occipital

Answer: A
Type: MC   Page Ref: 132
Topic: SKULL

7) The ethmoid bone does *not* contain:
A) olfactory foramina. B) crista galli.
C) cribriform plate. D) inferior nasal conchae.

Answer: D
Type: MC   Page Ref: 137
Topic: SKULL

8) Which of the following is *not* true for paranasal sinuses? They are:
A) lined by mucous membrane. B) chambers for voice resonance.
C) located in zygomatic bones. D) paired cavities.

Answer: C
Type: MC   Page Ref: 143–144
Topic: SKULL

9) The mandible has condylar processes and coronoid processes.
A) Both are articular in function.
B) Both are for muscle attachment.
C) The condylar process is articular; the coronoid process is for muscle attachment.
D) The condylar process is for muscle attachment; the coronoid process is articular.

Answer: C
Type: MC   Page Ref: 144
Topic: SKULL

10) Which of the following is *not* true?
A) The bodies of vertebrae form the posterior surface of the vertebral column.
B) The neural arch and body surround the spinal cord.
C) The spinous processes are directed posteriorly.
D) Laminae are posterior.

Answer: A
Type: MC   Page Ref: 150
Topic: VERTEBRAL COLUMN

11) Which of the following is true for the sacrum?
A) It is convex anteriorly.
B) The sacral canal is the extension of the vertebral canal.
C) Auricular surfaces articulate with the lumbar vertebrae.
D) It consists of seven fused vertebrae.

Answer: B
Type: MC   Page Ref: 155
Topic: VERTEBRAL COLUMN

12) The bone that consists of 5 fused bones, is triangular, and serves as one of the boundaries of the pelvic cavity is the:
A) coccyx. B) manubrium. C) coxal bone. D) sacrum.

Answer: D
Type: MC   Page Ref: 154
Topic: VERTEBRAL COLUMN

13) Which bone in the body articulates directly or indirectly with 22 other bones?
   A) sphenoid   B) sternum   C) sacrum   D) mandible

   Answer: B
   Type: MC   Page Ref: 158
   Topic: THORAX

14) Intercostal spaces:
   A) are filled by costal cartilages.
   B) contain muscles, blood vessels, and nerves.
   C) are grooves on the inner surfaces of ribs.
   D) contain bone marrow.

   Answer: B
   Type: MC   Page Ref: 160
   Topic: THORAX

TRUE/FALSE. Write 'T' if the statement is true and 'F' if the statement is false.

1) Phalanges of the digits are classed according to shape as short bones.

   Answer: FALSE
   Type: TF   Page Ref: 126
   Topic: TYPES OF BONES

2) Sesamoid and sutural bones are the two types of bones that are the most variable in number in the human body.

   Answer: TRUE
   Type: TF   Page Ref: 128
   Topic: TYPES OF BONES

3) The posterior (occipital) fontanel is at the junction of the lambdoid and sagittal sutures.

   Answer: TRUE
   Type: TF   Page Ref: 131, 148
   Topic: SKULL

4) The mandible articulates with the mandibular fossa of the occipital bone.

   Answer: FALSE
   Type: TF   Page Ref: 132, 144
   Topic: SKULL

5) The skull bone that articulates with all other cranial bones is the parietal bone.

   Answer: FALSE
   Type: TF   Page Ref: 135
   Topic: SKULL

6) The maxillae articulate with every facial bone except the mandible.

   Answer: TRUE
   Type: TF   Page Ref: 143
   Topic: SKULL

7)   The sacral ala articulates with the hipbones to form the sacroiliac joints.
     Answer: FALSE
     Type: TF   Page Ref: 155
     Topic: VERTEBRAL COLUMN

8)   The cervical curve is convex anteriorly.
     Answer: TRUE
     Type: TF   Page Ref: 150
     Topic: VERTEBRAL COLUMN

9)   The seventh cervical vertebra is called the vertebra prominens.
     Answer: FALSE
     Type: TF   Page Ref: 153
     Topic: VERTEBRAL COLUMN

10) The lumbar curve is concave anteriorly.
     Answer: FALSE
     Type: TF   Page Ref: 150
     Topic: VERTEBRAL COLUMN

11) The atlanto–axial joint allows for nodding the head to say yes.
     Answer: FALSE
     Type: TF   Page Ref: 152
     Topic: VERTEBRAL COLUMN

12) The three main parts of the sternum are manubrium, xiphoid process and cornua.
     Answer: FALSE
     Type: TF   Page Ref: 158
     Topic: THORAX

13) The rib is a flat bone that articulates at two points with a thoracic vertebra via its
     head and its tubercle.
     Answer: TRUE
     Type: TF   Page Ref: 126, 160
     Topic: THORAX

ESSAY.   Write your answer in the space provided or on a separate sheet of paper.

1)   List the bones of the vertebral column.
     Answer: 7 cervical, 12 thoracic, 5 lumbar, 5 sacral (fused), and usually 4 coccygeal
             (fused) vertebrae
     Type: ES   Page Ref: 150
     Topic: VERTEBRAL COLUMN

2) Name the seven processes that are found on most vertebrae and state the function of each.

   Answer: One spinous process and two transverse processes are for muscle attachment; two superior articular processes and two inferior articular processes articulate with adjacent vertebrae.
   Type: ES   Page Ref: 150
   Topic: VERTEBRAL COLUMN

3) Define and locate true ribs, false ribs, and floating ribs.

   Answer: True ribs (#1–7) attach directly to the sternum via their own costal cartilages. False ribs (#8–12) attach indirectly via costal cartilage #7, as for ribs #8–10, or do not attach anteriorly, as for ribs #11 and 12 (floating ribs).
   Type: ES   Page Ref: 159
   Topic: THORAX

4) Most ribs have four points at which they attach to other bones. Identify the parts of a rib referred to in the previous statement.

   Answer: Two facets on the head articulate with two adjacent vertebral bodies. The tubercle of a rib articulates with the transverse process of a thoracic vertebra. The anterior end of the rib attaches to a hyaline costal cartilage, which articulates with the sternum.
   Type: ES   Page Ref: 160
   Topic: THORAX

SHORT ANSWER. Write the word or phrase that best completes each statement or answers the question.

1) The musculoskeletal system consists of _____, _____, and _____.

   Answer: muscles, bones, joints
   Type: SA   Page Ref: 126
   Topic: DIVISIONS OF THE SKELETAL SYSTEM

2) The two divisions of the skeletal system are _____ and _____.

   Answer: axial, appendicular
   Type: SA   Page Ref: 126
   Topic: DIVISIONS OF THE SKELETAL SYSTEM

3) According to shape classification, the cuboid, hamate, and pisiform bones are _____ bones.

   Answer: short
   Type: SA   Page Ref: 126
   Topic: TYPES OF BONES

4) According to shape classification, the scapula is a/an _____ bone.

   Answer: flat
   Type: SA   Page Ref: 126
   Topic: TYPES OF BONES

5) If the squamous suture were separated (opened up), the parietal bone would be pulled away from the _____ bone.

Answer: temporal
Type: SA   Page Ref: 131
Topic: SKULL

6) The _____ bones of the skull are involved in each of the four main sutures.

Answer: parietal
Type: SA   Page Ref: 131
Topic: SKULL

7) Name the seven skull bones that help form each orbit.

Answer: frontal, ethmoid, sphenoid, lacrimal, palatine, zygomatic, maxilla
Type: SA   Page Ref: 147, 148
Topic: SKULL

8) Of the eight bones that form the cranium, only the _____ and _____ bones are paired bones.

Answer: temporal, parietal
Type: SA   Page Ref: 131, 132
Topic: SKULL

9) The _____ are convex surfaces on either side of the foramen magnum that articulate with the atlas.

Answer: occipital condyles
Type: SA   Page Ref: 134
Topic: SKULL

10) The two bones that form the zygomatic arch are the _____ and the _____.

Answer: temporal, zygomatic
Type: SA   Page Ref: 132
Topic: SKULL

11) The superior part of the nasal septum, the middle nasal conchae, and the crista galli are formed by the _____ bone.

Answer: ethmoid
Type: SA   Page Ref: 137
Topic: SKULL

12) Alveolar processes are features of two bones, the _____ and the _____,

Answer: maxillae, mandible
Type: SA   Page Ref: 143, 144
Topic: SKULL

13) The primary curves of the vertebral column are the _____ and the _____ curves. They are curved so that they are anteriorly _____.

Answer: thoracic, sacral, concave
Type: SA   Page Ref: 150
Topic: VERTEBRAL COLUMN

14) Paired spinal nerves pass through _____ foramina to carry information between the spinal cord and other parts of the body.

Answer: intervertebral
Type: SA   Page Ref: 150
Topic: VERTEBRAL COLUMN

15) The depressions on the superior border of the manubrium are called the _____ notch and the _____ notches.

Answer: suprasternal (jugular), clavicular
Type: SA   Page Ref: 158
Topic: THORAX

MATCHING. Choose the item in Column 2 that best matches each item in Column 1.

Match the skull bones in Column 1 with their features or descriptions in Column 2.
1) Column 1: nasal
   Column 2: forms part of the bridge of
             the nose

   Answer: forms part of the bridge of the nose
   Type: MA   Page Ref: 139
   Topic: SKULL

2) Column 1: vomer
   Column 2: inferior and posterior
             portion of nasal septum
   Foil: superior portion of nasal
             septum

   Answer: inferior and posterior portion of nasal septum
   Type: MA   Page Ref: 147
   Topic: SKULL

3) Column 1: maxilla
   Column 2: palatine process forms most
             of the hard palate
   Foil: lower jaw bone

   Answer: palatine process forms most of the hard palate
   Type: MA   Page Ref: 143
   Topic: SKULL

4) Column 1: temporal
   Column 2: contains the carotid and
             jugular foramina

   Answer: contains the carotid and jugular foramina
   Type: MA   Page Ref: 132
   Topic: SKULL

5)   Column 1: lacrimal
        Column 2: contains nasolacrimal ducts

        Answer: contains nasolacrimal ducts
        Type: MA   Page Ref: 144
        Topic: SKULL

6)   Column 1: palatine
        Column 2: horizontal plate forms the
                    posterior portion of the
                    hard palate

        Answer: horizontal plate forms the posterior portion of the hard palate
        Type: MA   Page Ref: 144
        Topic: SKULL

7)   Column 1: zygomatic
        Column 2: includes temporal process of
                    zygomatic arch

        Answer: includes temporal process of zygomatic arch
        Type: MA   Page Ref: 144
        Topic: SKULL

8)   Column 1: occipital
        Column 2: bears condyles that form
                    the atlanto-occipital joint

        Answer: bears condyles that form the atlanto-occipital joint
        Type: MA   Page Ref: 134
        Topic: SKULL

9)   Column 1: sphenoid
        Column 2: sella turcica protects the
                    pituitary gland

        Answer: sella turcica protects the pituitary gland
        Type: MA   Page Ref: 135
        Topic: SKULL

Match the vertebrae in Column 1 with their features or descriptions in Column 2.

10)   Column 1: cervical
        Column 2: small bodies
        Foil: four foramina

        Answer: small bodies
        Type: MA   Page Ref: 152
        Topic: VERTEBRAL COLUMN

11)   Column 1: cervical
        Column 2: contain transverse foramina

        Answer: contain transverse foramina
        Type: MA   Page Ref: 152
        Topic: VERTEBRAL COLUMN

12)   Column 1: thoracic
        Column 2: transverse processes
                articulate with tubercles of
                ribs
        Foil: articulate with costal
                cartilages

Answer: transverse processes articulate with tubercles of ribs
Type: MA   Page Ref: 153
Topic: VERTEBRAL COLUMN

13)   Column 1: thoracic
        Column 2: bodies have articular
                surfaces called facets and
                demifacets

Answer: bodies have articular surfaces called facets and demifacets
Type: MA   Page Ref: 153
Topic: VERTEBRAL COLUMN

14)   Column 1: lumbar
        Column 2: processes are thick and
                short
        Foil: articulate with the hip
                bone

Answer: processes are thick and short
Type: MA   Page Ref: 154
Topic: VERTEBRAL COLUMN

15)   Column 1: atlas
        Column 2: no body, no spinous
                process

Answer: no body, no spinous process
Type: MA   Page Ref: 152
Topic: VERTEBRAL COLUMN

16)   Column 1: axis
        Column 2: has a peglike process called
                a dens

Answer: has a peglike process called a dens
Type: MA   Page Ref: 152
Topic: VERTEBRAL COLUMN

Match the bone surface markings in Column 1 with their descriptions in Column 2.
17)   Column 1: crest
        Column 2: a prominent border or ridge

Answer: a prominent border or ridge
Type: MA   Page Ref: 130
Topic: BONE SURFACE MARKINGS

18) Column 1: tuberosity
Column 2: a large rounded, usually
roughened surface for
muscle attachment

Answer: a large rounded, usually roughened surface for muscle attachment
Type: MA   Page Ref: 130
Topic: TYPES OF BONES

19) Column 1: trochanter
Column 2: a large projection for
muscle attachment on the
femur

Answer: a large projection for muscle attachment on the femur
Type: MA   Page Ref: 130
Topic: BONE SURFACE MARKINGS

20) Column 1: facet
Column 2: a smooth, flat surface

Answer: a smooth, flat surface
Type: MA   Page Ref: 130
Topic: BONE SURFACE MARKINGS

21) Column 1: condyle
Column 2: a rounded articular surface

Answer: a rounded articular surface
Type: MA   Page Ref: 130
Topic: BONE SURFACE MARKINGS

22) Column 1: epicondyle
Column 2: a prominence above a
condyle

Answer: a prominence above a condyle
Type: MA   Page Ref: 130
Topic: BONE SURFACE MARKINGS

23) Column 1: meatus
Column 2: a tubelike passageway

Answer: a tubelike passageway
Type: MA   Page Ref: 130
Topic: BONE SURFACE MARKINGS

24) Column 1: fossa
Column 2: a depression

Answer: a depression
Type: MA   Page Ref: 130
Topic: BONE SURFACE MARKINGS

CHAPTER 7 The Skeletal System: The Appendicular Skeleton

MULTIPLE CHOICE. Choose the one alternative that best completes the statement or answers the question.

1) The coracoid process of the scapula is:
   A) a site of muscle attachment.   B) an extension of the spine.
   C) the high point of the shoulder.   D) all of the above.

   Answer: A
   Type: MC   Page Ref: 168
   Topic: PECTORAL GIRDLE

2) The scapula:
   A) articulates with the clavicle and the humerus.
   B) is a flat bone covering the posterior regions of ribs 5-10.
   C) has a superior angle that forms the high point of the shoulder.
   D) all of the above.

   Answer: A
   Type: MC   Page Ref: 168
   Topic: PECTORAL GIRDLE

3) Which of the following is true?
   A) The medial end of the clavicle is the sternal extremity.
   B) The curvature of the medial one-third of the clavicle is concave anteriorly.
   C) The clavicle articulates medially with the body of the sternum and laterally with
      the acromion of the scapula.
   D) All of the above statements are true.

   Answer: B
   Type: MC   Page Ref: 168
   Topic: PECTORAL GIRDLE

4) The distal end of the humerus has all of the following features *except*:
   A) capitulum.   B) trochlea.
   C) olecranon.   D) medial and lateral epicondyles.

   Answer: C
   Type: MC   Page Ref: 170
   Topic: UPPER LIMB

5) Which of the following is found on the anterior surface of the humerus?
   A) deltoid tuberosity   B) radial fossa
   C) greater tubercle   D) medial epicondyle

   Answer: B
   Type: MC   Page Ref: 170
   Topic: UPPER LIMB

6)   The ulna:
      A) articulates with the trochlea of the humerus.
      B) is the lateral bone of the forearm.
      C) is the shorter bone of the forearm.
      D) has a proximal disc-shaped head.

      Answer: A
      Type: MC   Page Ref: 170, 171
      Topic: UPPER LIMB

7)   Which of the following is true of the radius?
      A) It is the medial bone of the forearm.
      B) Its head articulates with the distal end of the ulna.
      C) It articulates with the trochlear notch of the humerus.
      D) None of the above.

      Answer: D
      Type: MC   Page Ref: 171-173
      Topic: UPPER LIMB

8)   The bony landmarks commonly referred to as "knuckles" are the:
      A) bases of proximal phalanges.   B) heads of proximal phalanges.
      C) bases of metacarpals.   D) heads of metacarpals.

      Answer: D
      Type: MC   Page Ref: 175
      Topic: UPPER LIMB

9)   Which of these does *not* belong to the distal row of carpal bones?
      A) pisiform   B) trapezoid   C) trapezium   D) capitate

      Answer: A
      Type: MC   Page Ref: 174
      Topic: UPPER LIMB

10)   Which of the following is *not* part of the ilium?
      A) greater sciatic notch   B) inferior gluteal line
      C) lesser sciatic notch   D) auricular surface

      Answer: C
      Type: MC   Page Ref: 177
      Topic: PELVIC GIRDLE

11)   The cup-shaped depression formed by all three portions of the coxal bone is the
      _____.
      A) false pelvis   B) obturator foramen
      C) acetabulum   D) glenoid cavity

      Answer: C
      Type: BI   Page Ref: 177
      Topic: PELVIC GIRDLE

12) The male skeleton, as compared to the female:
A) has heavier bones.
B) has rougher, larger tuberosities and ridges.
C) has a heart-shaped pelvic inlet.
D) all of the above.

Answer: D
Type: MC   Page Ref: 178, 180, 186
Topic: PELVIC GIRDLE

13) The patellar surface, with which the patella articulates, is located on the _____ end of the _____.
A) proximal, tibia   B) distal, tibia   C) proximal, femur   D) distal, femur

Answer: D
Type: BI   Page Ref: 180
Topic: LOWER LIMB

14) Which of the following is an articular feature of the tibia?
A) tibial tuberosity   B) anterior border
C) lateral condyle   D) none of the above

Answer: C
Type: MC   Page Ref: 181
Topic: LOWER LIMB

15) Which of the following is *not* a tarsal bone?
A) navicular   B) hamate   C) cuboid   D) cuneiform

Answer: B
Type: MC   Page Ref: 183
Topic: LOWER LIMB

16) There are a total of _____ phalanges in the human body.
A) 14   B) 20   C) 28   D) 56

Answer: D
Type: MC   Page Ref: 175, 183
Topic: LOWER LIMB

17) The largest of the tarsal bones is the:
A) talus.   B) calcaneus.   C) cuboid.   D) navicular.

Answer: B
Type: MC   Page Ref: 183
Topic: LOWER LIMB

18) The arches of the foot:
A) are four in number.
B) are formed by metatarsal and tarsal bones.
C) are rigid.
D) all of the above.

Answer: B
Type: MC   Page Ref: 186
Topic: LOWER LIMB

TRUE/FALSE. Write 'T' if the statement is true and 'F' if the statement is false.

1) The inferior surface of the clavicle is rougher than the superior surface, due, in part, to the presence of the conoid tubercle and the costal tuberosity.

Answer: TRUE
Type: TF   Page Ref: 168
Topic: PECTORAL GIRDLE

2) The anatomical neck of the humerus is proximal to the surgical neck.

Answer: TRUE
Type: TF   Page Ref: 170
Topic: UPPER LIMB

3) The capitulum of the humerus is lateral to the trochlea.

Answer: TRUE
Type: TF   Page Ref: 170
Topic: UPPER LIMB

4) The head of the ulna is on the distal end of the bone.

Answer: TRUE
Type: TF   Page Ref: 171
Topic: UPPER LIMB

5) The raised roughened area, just distal to the neck of the radius, that serves as attachment for the biceps muscle is the biceps tuberosity.

Answer: FALSE
Type: TF   Page Ref: 173
Topic: UPPER LIMB

6) The posterior inferior portion of the coxal bone is the pubis.

Answer: FALSE
Type: TF   Page Ref: 175
Topic: PELVIC GIRDLE

7) The false pelvis is superior to the true pelvis.

Answer: TRUE
Type: TF   Page Ref: 177
Topic: PELVIC GIRDLE

8) The intertrochanteric line is near the proximal end of the femur, on the anterior surface.

Answer: TRUE
Type: TF   Page Ref: 180
Topic: LOWER LIMB

9) The fibula, the lateral bone of the leg, is a weight-bearing bone.

Answer: FALSE
Type: TF   Page Ref: 183
Topic: LOWER LIMB

ESSAY. Write your answer in the space provided or on a separate sheet of paper.

1) List the bones that make up the appendicular skeleton under the headings pectoral girdles, upper extremity, pelvic girdle, lower extremity.

Answer: Pectoral girdle: 2 scapulae, 2 clavicles
Upper extremity: humerus, ulna, radius, 8 carpal bones, 5 metacarpals, 14 phalanges
Pelvic girdle: 2 coxal bones
Lower extremity: femur, patella, tibia, fibula, 7 tarsal bones, 5 metatarsals, 14 phalanges

Type: ES   Page Ref: 167–183
Topic: APPENDICULAR SKELETON

2) Give an anatomical description of the scapula, describing the features that are visible from the posterior aspect.

Answer: Should include a description of borders, angles, spine, acromion, supraspinous fossa, infraspinous fossa, scapular notch.

Type: ES   Page Ref: 168, 169
Topic: PECTORAL GIRDLE

3) Name the articular surfaces of the coxal bones and identify the bones that articulate at each surface.

Answer: Auricular surface: ilium articulates with sacrum
Pubic symphysis: 2 pubic bones articulate with each other
Acetabulum: ilium, ischium, and pubis articulate with femur

Type: ES   Page Ref: 177
Topic: PELVIC GIRDLE

4) Give an anatomical description of the femur as it would appear from the posterior.

Answer: Features described: head, neck, greater and lesser trochanters, intertrochanteric crest, linea aspera, gluteal tuberosity, shaft, medial and lateral condyles, intercondylar fossa, medial and lateral epicondyles.

Type: ES   Page Ref: 180, 182
Topic: LOWER LIMB

SHORT ANSWER. Write the word or phrase that best completes each statement or answers the question.

1) The greater tubercle of the humerus is separated from the lesser tubercle by the

_____.

Answer: intertubercular sulcus
Type: SA   Page Ref: 170
Topic: UPPER LIMB

2) The knuckle joints are formed by the _____ and the _____.

Answer: metacarpals, phalanges
Type: SA   Page Ref: 175
Topic: UPPER LIMB

3) The projection at the proximal end of the ulna that forms the prominence of the elbow is the _____.
Answer: olecranon
Type: SA   Page Ref: 170
Topic: UPPER LIMB

4) The head of the radius fits into the _____ notch of the ulna.
Answer: radial
Type: SA   Page Ref: 171
Topic: UPPER LIMB

5) The ulna and radius each bear a _____ process at their distal ends.
Answer: styloid
Type: SA   Page Ref: 171, 172
Topic: UPPER LIMB

6) Which carpal bone in the proximal row has three articular surfaces?
Answer: triquetrum
Type: SA   Page Ref: 174
Topic: UPPER LIMB

7) The carpal bone with a hook-shaped process on its anterior surface is the _____.
Answer: hamate
Type: SA   Page Ref: 175
Topic: UPPER LIMB

8) A large depression on the _____ surface of the ilium is called the iliac fossa.
Answer: medial
Type: SA   Page Ref: 188
Topic: PELVIC GIRDLE

9) The coxal bones unite anteriorly with each other to form a joint called the _____. They unite posteriorly with the _____.
Answer: pubic symphysis, sacrum
Type: SA   Page Ref: 175
Topic: PELVIC GIRDLE

10) The distal end of the fibula forms the _____ malleolus.
Answer: lateral
Type: SA   Page Ref: 183
Topic: LOWER LIMB

11) The fibular notch is located at the distal end of the _____.
Answer: tibia
Type: SA   Page Ref: 183
Topic: LOWER LIMB

12) The distal end of the tibia and fibula articulate with each other and with the _____.
Answer: talus
Type: SA   Page Ref: 183
Topic: LOWER LIMB

13) The projection on the superior surface of the tibia that separates the two condyles is the _____.
Answer: intercondylar eminence
Type: SA   Page Ref: 181
Topic: LOWER LIMB

MATCHING.   Choose the item in Column 2 that best matches each item in Column 1.

Match the bones of the appendicular skeleton in Column 1 with their features or descriptions in Column 2.

1)   Column 1: clavicle
     Column 2: articulates with acromion
                   and manubrium
     Foil: lateral half is convex
                   anteriorly
     Answer: articulates with acromion and manubrium
     Type: MA   Page Ref: 168
     Topic: APPENDICULAR SKELETON

2)   Column 1: scapula
     Column 2: has three large nonarticular
                   fossae that occupy a large
                   portion of the surface of
                   the bone
     Foil: articulates with thoracic
                   region of vertebral column
     Answer: has three large nonarticular fossae that occupy a large portion of the surface of
               the bone
     Type: MA   Page Ref: 168
     Topic: APPENDICULAR SKELETON

3)   Column 1: ulna
     Column 2: radial notch on lateral
                   surface
     Answer: radial notch on lateral surface
     Type: MA   Page Ref: 171
     Topic: APPENDICULAR SKELETON

4)   Column 1: radius
     Column 2: lateral bone of forearm
     Answer: lateral bone of forearm
     Type: MA   Page Ref: 171
     Topic: APPENDICULAR SKELETON

5)   Column 1: middle phalanx
        Column 2: absent in pollex
        Foil: found only in digit 3

        Answer: absent in pollex
        Type: MA   Page Ref: 175
        Topic: APPENDICULAR SKELETON

6)   Column 1: ilium
        Column 2: articulates with the sacrum

        Answer: articulates with the sacrum
        Type: MA   Page Ref: 177
        Topic: APPENDICULAR SKELETON

7)   Column 1: pubis
        Column 2: inferior anterior portion of
                  hipbone

        Answer: inferior anterior portion of hipbone
        Type: MA   Page Ref: 177
        Topic: APPENDICULAR SKELETON

8)   Column 1: femur
        Column 2: greater and lesser
                  trochanters at proximal end
        Foil: intercondylar fossa at
                    proximal end

        Answer: greater and lesser trochanters at proximal end
        Type: MA   Page Ref: 180
        Topic: APPENDICULAR SKELETON

9)   Column 1: patella
        Column 2: completely enclosed in a
                  tendon

        Answer: completely enclosed in a tendon
        Type: MA   Page Ref: 180
        Topic: APPENDICULAR SKELETON

10)   Column 1: tibia
        Column 2: medial malleolus at distal
                  end
        Foil: lateral bone of leg

        Answer: medial malleolus at distal end
        Type: MA   Page Ref: 183
        Topic: APPENDICULAR SKELETON

11) Column 1: talus
Column 2: articulates with tibia and fibula
Foil: smallest tarsal bone

Answer: articulates with tibia and fibula
Type: MA   Page Ref: 183
Topic: APPENDICULAR SKELETON

12) Column 1: metatarsal
Column 2: base articulates with a tarsal bone

Answer: base articulates with a tarsal bone
Type: MA   Page Ref: 183
Topic: APPENDICULAR SKELETON

13) Column 1: proximal phalanx
Column 2: base forms joint with a metatarsal

Answer: base forms joint with a metatarsal
Type: MA   Page Ref: 183
Topic: APPENDICULAR SKELETON

Match the common names of bones in Column 1 with their anatomical names in Column 2.

14) Column 1: collarbone
Column 2: clavicle

Answer: clavicle
Type: MA   Page Ref: 167
Topic: APPENDICULAR SKELETON

15) Column 1: thighbone
Column 2: femur

Answer: femur
Type: MA   Page Ref: 180
Topic: APPENDICULAR SKELETON

16) Column 1: heel bone
Column 2: calcaneus
Foil: talus

Answer: calcaneus
Type: MA   Page Ref: 183
Topic: APPENDICULAR SKELETON

17) Column 1: thumb
Column 2: pollex

Answer: pollex
Type: MA   Page Ref: 175
Topic: APPENDICULAR SKELETON

18)   Column 1: arm
        Column 2: humerus
        Foil: ulna

        Answer: humerus
        Type: MA   Page Ref: 170
        Topic: APPENDICULAR SKELETON

19)   Column 1: big toe
        Column 2: hallux

        Answer: hallux
        Type: MA   Page Ref: 183
        Topic: APPENDICULAR SKELETON

20)   Column 1: shinbone
        Column 2: tibia
        Foil: fibula

        Answer: tibia
        Type: MA   Page Ref: 181
        Topic: APPENDICULAR SKELETON

21)   Column 1: shoulder blade
        Column 2: scapula

        Answer: scapula
        Type: MA   Page Ref: 168
        Topic: APPENDICULAR SKELETON

CHAPTER 8   Joints

MULTIPLE CHOICE. Choose the one alternative that best completes the statement or answers the question.

1)   Joints may be classified according to the amount of movement permitted. The following factors affect joint movement, *except* for:
A) tension of ligaments.
B) shape of articulating surfaces.
C) arrangement of ligaments, tendons, and muscles.
D) length of the bones.

Answer: D
Type: MC   Page Ref: 190
Topic: CLASSIFICATION OF JOINTS

2)   Which of the following pairs of terms is most closely associated or matched?
A) atlanto–occipital, amphiarthrosis   B) vertebrocostal, synarthrosis
C) talocrural, diarthrosis   D) sacroiliac, synarthrosis

Answer: C
Type: MC   Page Ref: 219
Topic: CLASSIFICATION OF JOINTS

3)   If a joint is enclosed in a tough connective tissue capsule and if it contains a joint cavity, it is classified as:
A) synovial.   B) fibrous.
C) cartilaginous.   D) all of the above.

Answer: A
Type: MC   Page Ref: 191
Topic: CLASSIFICATION OF JOINTS

4)   Which type of joint has the most movement?
A) synarthrosis   B) diarthrosis   C) gomphosis   D) amphiarthrosis

Answer: B
Type: MC   Page Ref: 190
Topic: CLASSIFICATION OF JOINTS

5)   Fibrous connective tissue firmly holds the articular surfaces of bones together in:
A) fibrous joints.   B) cartilaginous joints.
C) synovial joints.   D) all of the above.

Answer: A
Type: MC   Page Ref: 190
Topic: FIBROUS JOINTS

6)   A fibrous joint may be a:
A) suture.   B) gomphosis.
C) syndesmosis.   D) all of the above.

Answer: D
Type: MC   Page Ref: 190
Topic: FIBROUS JOINTS

7)   A _____ is a type of joint in which two bones are held together by a disc of
      fibrocartilage.
      A) symphysis   B) synchondrosis   C) suture   D) synovial joint
      Answer: A
      Type: BI   Page Ref: 191
      Topic: CARTILAGINOUS JOINTS

8)   Epiphyseal plates of growing bones may be classified as:
      A) cartilaginous joints.   B) synarthrotic joints.
      C) synchondroses.   D) all of the above.
      Answer: D
      Type: MC   Page Ref: 191
      Topic: CARTILAGINOUS JOINTS

9)   The articular cartilage:
      1. consists of two layers; an outer layer of dense fibrous connective tissue and an
      inner layer of loose connective tissue.
      2. firmly unites the articulating bones.
      3. shields the surfaces of the articulating bones at a synovial joint.
      4. helps absorb shock and reduces friction at a synovial joint.
      A) 1, 2, 3   B) 2, 3, 4   C) 3, 4   D) 1, 2
      Answer: C
      Type: MC   Page Ref: 191
      Topic: SYNOVIAL JOINTS

10)   A synovial membrane contains:
      A) dense irregular connective tissue.
      B) a large quantity of collagen fibers in bundles.
      C) simple cuboidal epithelium that secretes synovial fluid.
      D) loose connective tissue.
      Answer: D
      Type: MC   Page Ref: 191
      Topic: SYNOVIAL JOINTS

11)   Structures that are most responsible for holding bones together at a synovial joint
      are:
      A) tendons.   B) articular cartilages.
      C) synovial membranes.   D) ligaments.
      Answer: D
      Type: MC   Page Ref: 191
      Topic: SYNOVIAL JOINTS

12)   Articular discs of synovial joints:
      A) are pads of hyaline cartilage.
      B) move freely within the joint cavity.
      C) are found in the space between the ends of the bones.
      D) are found in all synovial joints.
      Answer: C
      Type: MC   Page Ref: 192
      Topic: SYNOVIAL JOINTS

13) The surfaces of the bones at a gliding joint perform the movement(s) of:
A) side-to-side movement.   B) rotation.
C) flexion and extension.   D) all of the above.

Answer: A
Type: MC   Page Ref: 194
Topic: SYNOVIAL JOINTS

14) Which of the following terms could describe a joint at which flexion and extension are the *only* movements?
1. pivot joint
2. hinge joint
3. monaxial joint
4. biaxial joint
A) 1, 3   B) 2, 3   C) 1, 4   D) 2, 4

Answer: B
Type: MC   Page Ref: 194, 195
Topic: SYNOVIAL JOINTS

15) The action that moves the palm of the hand into anatomical position is:
A) pronation.   B) supination.   C) inversion.   D) eversion.

Answer: B
Type: MC   Page Ref: 200
Topic: SYNOVIAL JOINTS

16) The articular capsule of the hip joint is one of the strongest in the body. It consists partially of:
A) transverse humeral ligament.   B) pubofemoral ligament.
C) glenoid labrum.   D) all of the above.

Answer: B
Type: MC   Page Ref: 211
Topic: SYNOVIAL JOINTS

17) The primary type of movement permitted at a pivot joint is:
A) rotation.   B) circumduction.
C) abduction and adduction.   D) all of the above.

Answer: A
Type: MC   Page Ref: 200
Topic: SYNOVIAL JOINTS

TRUE/FALSE. Write 'T' if the statement is true and 'F' if the statement is false.

1) The joint between two bodies of adjacent vertebrae is a synchondrosis.
Answer: FALSE
Type: TF   Page Ref: 191
Topic: CLASSIFICATION OF JOINTS

2) All fibrous joints are synarthrotic; that is, they are immovable.
Answer: FALSE
Type: TF   Page Ref: 190
Topic: FIBROUS JOINTS

3) The joints between the true ribs and the sternum are syndesmoses.

Answer: FALSE
Type: TF   Page Ref: 191
Topic: CARTILAGINOUS JOINTS

4) Synovial fluid is a somewhat viscous liquid that contains hyaluronic acid and fluid formed from blood plasma.

Answer: TRUE
Type: TF   Page Ref: 191
Topic: SYNOVIAL JOINTS

5) Ligaments at synovial joints may be extracapsular or intracapsular.

Answer: TRUE
Type: TF   Page Ref: 192
Topic: SYNOVIAL JOINTS

6) The shoulder joint is an example of a triaxial joint.

Answer: TRUE
Type: TF   Page Ref: 194
Topic: SYNOVIAL JOINTS

7) Bursae are connective tissue sacs that are responsible for cushioning the movement of structures at joints.

Answer: TRUE
Type: TF   Page Ref: 193
Topic: SYNOVIAL JOINTS

8) Interphalangeal joints are synovial joints at which flexion, extension, and hyperextension occur.

Answer: FALSE
Type: TF   Page Ref: 195
Topic: SYNOVIAL JOINTS

9) Abduction occurs when a bone moves toward the midline.

Answer: FALSE
Type: TF   Page Ref: 195
Topic: SYNOVIAL JOINTS

10) Bending the ankle so that the foot moves downward is the movement called plantar flexion.

Answer: TRUE
Type: TF   Page Ref: 200
Topic: SYNOVIAL JOINTS

11) Medial and lateral menisci are found in the joint cavity of the elbow.

Answer: FALSE
Type: TF   Page Ref: 209
Topic: SYNOVIAL JOINTS

ESSAY. Write your answer in the space provided or on a separate sheet of paper.

1) Define the terms synarthrosis, amphiarthrosis, and diarthrosis and give an example of each.

Answer: Synarthrosis: immovable joint (e.g., epiphyseal plate or skull sutures)
Amphiarthrosis: slightly movable joint (e.g., pubic symphysis or interosseus ligament between radius and ulna)
Diarthrosis: fully movable joint (e.g., jaw, elbow, knee joints, etc.)
Type: ES   Page Ref: 190
Topic: CLASSIFICATION OF JOINTS

2) Describe the following features of a synovial joint and state the function of each: articular cartilage, fibrous capsule, synovial membrane, synovial fluid.

Answer: Answer: see pp. 191, 192
Type: ES   Page Ref: 191, 192
Topic: SYNOVIAL JOINTS

3) Give the structural and functional classification, list the possible movements, and name the articular surfaces of the bones for a) the shoulder joint, b) the metacarpophalangeal joints, c) the hip joint.

Answer: a) The shoulder joint is between the head of the humerus and the glenoid cavity of the scapula, and is synovial and diarthrotic. Movements are flexion, extension, hyperextension, abduction, adduction, medial and lateral rotation, and circumduction of the arm.
b) The metacarpophalangeal joints are between the heads of the metacarpals and bases of the proximal phalanges, and are synovial and diarthrotic. Movements are flexion, extension, abduction, adduction, and circumduction of the proximal phalanges.
c) The hip joint is between the head of the femur and the acetabulum of the coxal bone, and is synovial and diarthrotic. Movements are flexion, extension, abduction, adduction, circumduction, and rotation of the thigh.
Type: ES   Page Ref: 218, 219
Topic: SELECTED JOINTS OF THE BODY

SHORT ANSWER. Write the word or phrase that best completes each statement or answers the question.

1) The general anatomical term for the regions of contact between bones of the appendicular skeleton, at which various amounts of movements occur, is _____.

Answer: articulations
Type: SA   Page Ref: 190
Topic: CLASSIFICATION OF JOINTS

2) In order to fully understand kinesiology (study of movement), one would be advised to study _____ (scientific study of joints) first.

Answer: arthrology
Type: SA   Page Ref: 190
Topic: CLASSIFICATION OF JOINTS

3) The distal tibiofibular joint, which allows some movement, is classified structurally as a _____.
Answer: syndesmosis
Type: SA   Page Ref: 191
Topic: FIBROUS JOINTS

4) The name of the joint type where a tooth fits into an alveolar socket is _____.
Answer: gomphosis
Type: SA   Page Ref: 191
Topic: FIBROUS JOINTS

5) A symphysis such as the pubic symphysis allows a small amount of movement, and is functionally classed as a/an _____.
Answer: amphiarthrosis
Type: SA   Page Ref: 191
Topic: CARTILAGINOUS JOINTS

6) The articulating portions of a joint, such as articular cartilage, receive their nourishment from the _____.
Answer: synovial fluid
Type: SA   Page Ref: 191
Topic: SYNOVIAL JOINTS

7) A structure that consists of parallel bundles of collagen fibers and that is part of the capsule of some synovial joints is called a/an _____.
Answer: ligament or intracapsular ligament
Type: SA   Page Ref: 192
Topic: SYNOVIAL JOINTS

8) The movement of the jaw in a forward direction parallel to the ground is called _____.
Answer: protraction
Type: SA   Page Ref: 199
Topic: SYNOVIAL JOINTS

9) The knee joint, the largest joint, is actually three joints: two tibiofemoral joints and one _____ joint.
Answer: patellofemoral
Type: SA   Page Ref: 214
Topic: SYNOVIAL JOINTS

MATCHING.   Choose the item in Column 2 that best matches each item in Column 1.

Match the types of synovial joints in Column 1 with the descriptions in Column 2.
1) Column 1: gliding (arthrodial)
Column 2: flat articulating surfaces

Answer: flat articulating surfaces
Type: MA   Page Ref: 201
Topic: SYNOVIAL JOINTS

2)   Column 1: hinge (ginglymus)
        Column 2: one articulating surface is
                convex and fits into the
                other, which is concave

        Answer: one articulating surface is convex and fits into the other, which is concave
        Type: MA   Page Ref: 201
        Topic: SYNOVIAL JOINTS

3)   Column 1: condyloid (ellipsoidal)
        Column 2: oval–shaped projection fits
                into an ellipsoidal
                depression
        Foil: one bone articulates with
                another like a rider in a
                saddle

        Answer: oval–shaped projection fits into an ellipsoidal depression
        Type: MA   Page Ref: 201
        Topic: SYNOVIAL JOINTS

4)   Column 1: ball–and–socket (spheroid)
        Column 2: the only synovial joint that
                allows triaxial movement
        Foil: a bone rotates around its
                long axis as it articulates
                within a ring of bone and
                ligament

        Answer: the only synovial joint that allows triaxial movement
        Type: MA   Page Ref: 201
        Topic: SYNOVIAL JOINTS

Match the types of synovial joints in Column 1 with the examples in
Column 2.
5)   Column 1: gliding
        Column 2: between tarsal bones

        Answer: between tarsal bones
        Type: MA   Page Ref: 200
        Topic: SYNOVIAL JOINTS

6)   Column 1: hinge
        Column 2: knee joint
        Foil: hip joint

        Answer: knee joint
        Type: MA   Page Ref: 200
        Topic: SYNOVIAL JOINTS

7)   Column 1: pivot
        Column 2: between atlas and dens of
                  axis

        Answer: between atlas and dens of axis
        Type: MA   Page Ref: 200
        Topic: SYNOVIAL JOINTS

8)   Column 1: saddle
        Column 2: between trapezium and first
                  metacarpal
        Foil: wrist joint

        Answer: between trapezium and first metacarpal
        Type: MA   Page Ref: 200
        Topic: SYNOVIAL JOINTS

Match the special movements in Column 1 with the joints in Column 2.
9)   Column 1: elevation and depression
        Column 2: acromioclavicular
        Foil: knee

        Answer: acromioclavicular
        Type: MA   Page Ref: 199
        Topic: SYNOVIAL JOINTS

10)   Column 1: inversion and eversion
        Column 2: intertarsal
        Foil: interphalangeal

        Answer: intertarsal
        Type: MA   Page Ref: 199
        Topic: SYNOVIAL JOINTS

11)   Column 1: dorsiflexion and plantar
                  flexion
        Column 2: talocrural

        Answer: talocrural
        Type: MA   Page Ref: 199, 200
        Topic: SYNOVIAL JOINTS

12)   Column 1: pronation and supination
        Column 2: radioulnar
        Foil: wrist

        Answer: radioulnar
        Type: MA   Page Ref: 200
        Topic: SYNOVIAL JOINTS

CHAPTER 9   Muscle Tissue

MULTIPLE CHOICE. Choose the one alternative that best completes the statement or answers the question.

1)   The ability of muscle tissue to shorten and thicken is:
      A) elasticity.   B) extensibility.   C) stability.   D) contractility.

      Answer: D
      Type: BI   Page Ref: 233
      Topic: CHARACTERISTICS OF MUSCLE TISSUE

2)   The ability of muscle tissue to be stretched without being damaged is:
      A) elasticity.   B) extensibility.   C) excitability.   D) contractility.

      Answer: B
      Type: BI   Page Ref: 226
      Topic: CHARACTERISTICS OF MUSCLE TISSUE

3)   The function of blood vessels in skeletal muscle are:
      1. delivery of oxygen and nutrients to the muscle
      2. delivery of ATP to the muscle
      3. stimulation of muscle cells
      4. delivery of heat to the muscle
      5. removal of wastes and heat
      A) 1, 2, 3, 4   B) 2, 3, 4   C) 1, 2, 5   D) 1 and 5 only

      Answer: D
      Type: MC   Page Ref: 227
      Topic: SKELETAL MUSCLE TISSUE

4)   Deep fascia is a sheet of connective tissue that:
      A) is composed of areolar connective tissue.
      B) is composed of dense irregular connective tissue.
      C) is located in the subcutaneous tissue.
      D) none of the above.

      Answer: B
      Type: MC   Page Ref: 226
      Topic: SKELETAL MUSCLE TISSUE

5)   Place the following events in the correct order.
      1. ACh is released into the synaptic cleft.
      2. Synaptic vesicles fuse with the plasma membrane.
      3. A muscle action potential is triggered.
      4. An action potential arrives at the axon terminal.
      5. ACh binds to integral protein receptors of motor end plate.
      A) 2, 4, 5, 1, 3   B) 4, 3, 1, 2, 5   C) 1, 2, 4, 3, 5   D) 4, 2, 1, 5, 3

      Answer: D
      Type: MC   Page Ref: 229
      Topic: NEUROMUSCULAR JUNCTION

6)   The three types of protein strands found in myofibrils of skeletal muscle are:
      A) thick, thin, and elastic.   B) anisotropic, isotropic, and striated.
      C) troponin, tropomyosin, and actin.   D) titin, isotropic, and elastic.

      Answer: A
      Type: MC   Page Ref: 229, 230
      Topic: SKELETAL MUSCLE TISSUE

7)   Which of the following is true of myosin?
      1. It is located in the A band of the sarcomere.
      2. It binds to tropomyosin during contraction.
      3. It forms thick filaments.
      4. The molecules are helix shaped.
      A) 1, 2, 3, 4   B) 1, 2, 3   C) 1, 3   D) 2, 4

      Answer: C
      Type: MC   Page Ref: 231
      Topic: SKELETAL MUSCLE TISSUE

8)   In skeletal muscle, the sarcoplasmic reticulum:
      A) forms transverse (T) tubules.   B) is similar to the Golgi apparatus.
      C) contains extracellular fluid.   D) releases Ca^{2+} to trigger contraction.

      Answer: D
      Type: MC   Page Ref: 234
      Topic: SKELETAL MUSCLE TISSUE

9)   The close association of sarcoplasmic reticulum and transverse (T) tubules is called a:
      A) sarcomere.   B) triad.   C) cistern.   D) cross bridge.

      Answer: B
      Type: BI   Page Ref: 234
      Topic: SKELETAL MUSCLE TISSUE

10)   During contraction in a skeletal muscle fiber:
      A) the thick filaments meet at the center of the sarcomere.
      B) the sarcomere length does not change.
      C) actin and myosin molecules contract, causing filaments to shorten.
      D) myosin cross bridges move the thin filaments so that their ends meet or overlap
         in the center of the sarcomere.

      Answer: D
      Type: MC   Page Ref: 235
      Topic: SKELETAL MUSCLE TISSUE

11)   An increase (hypertrophy) or decrease (atrophy) in muscle size is due to:
      A) increased or decreased numbers of muscle cells.
      B) increased or decreased deposition of adipose tissue.
      C) increased or decreased numbers of myofibrils and organelles.
      D) both A and C.

      Answer: C
      Type: MC   Page Ref: 242
      Topic: SKELETAL MUSCLE TISSUE

12) The skeletal muscle fibers that fatigue most easily are:
   A) red muscle fibers.   B) slow fibers.
   C) fast fibers.   D) intermediate fibers.

   Answer: C
   Type: MC   Page Ref: 236
   Topic: TYPES OF SKELETAL MUSCLE FIBERS

13) Compared to skeletal muscle, cardiac muscle differs in that it:
   A) contracts for shorter periods of time.
   B) has a short refractory period.
   C) has a different arrangement of thick and thin filaments.
   D) has branched cells.

   Answer: D
   Type: MC   Page Ref: 238
   Topic: CARDIAC MUSCLE TISSUE

14) The two types of smooth muscle are:
   A) visceral and parietal.   B) red and white.
   C) visceral and multiunit.   D) radial and circular.

   Answer: C
   Type: BI   Page Ref: 240
   Topic: SMOOTH MUSCLE TISSUE

15) As skeletal muscle ages, after age 30, which of the following does *not* occur?
   A) decrease in maximal strength
   B) decrease in the relative number of slow fibers
   C) replacement of muscle by fat
   D) slowing of muscle reflexes

   Answer: B
   Type: MC   Page Ref: 243
   Topic: AGING

TRUE/FALSE. Write 'T' if the statement is true and 'F' if the statement is false.

1) A motor neuron is a nerve cell that stimulates a muscle cell, causing a contraction.

   Answer: TRUE
   Type: TF   Page Ref: 227
   Topic: SKELETAL MUSCLE TISSUE

2) A bundle (fascicle) of skeletal muscle cells is separated from neighboring bundles by perimysium.

   Answer: TRUE
   Type: TF   Page Ref: 226
   Topic: SKELETAL MUSCLE TISSUE

3) Epimysium surrounds skeletal muscle cells within fascicles.

   Answer: FALSE
   Type: TF   Page Ref: 226
   Topic: SKELETAL MUSCLE TISSUE

4) A motor neuron together with all the skeletal muscle fibers it stimulates is called a motor unit.

Answer: TRUE
Type: TF   Page Ref: 227
Topic: MOTOR UNIT

5) The darker area in a sarcomere is the I band.

Answer: FALSE
Type: TF   Page Ref: 231
Topic: SKELETAL MUSCLE TISSUE

6) The contractile elements of skeletal muscle cells are myofibrils.

Answer: TRUE
Type: TF   Page Ref: 229
Topic: SKELETAL MUSCLE TISSUE

7) Cardiac muscle cells contain more mitochondria than do skeletal muscle cells.

Answer: TRUE
Type: TF   Page Ref: 238
Topic: CARDIAC MUSCLE TISSUE

8) Cardiac muscle tissue may be classified as either visceral or multiunit smooth muscle.

Answer: FALSE
Type: TF   Page Ref: 240
Topic: SMOOTH MUSCLE TISSUE

9) New skeletal muscle cells can arise from satellite cells.

Answer: TRUE
Type: TF   Page Ref: 241
Topic: REGENERATION

10) The specialized region of sarcolemma that is located at a neuromuscular junction is called the synaptic cleft.

Answer: FALSE
Type: TF   Page Ref: 228
Topic: NEUROMUSCULAR JUNCTION

ESSAY. Write your answer in the space provided or on a separate sheet of paper.

1) List four functions of muscle tissue and in one sentence for each function, elaborate on the function or give an example to illustrate the function.

Answer: 1. Motion: movement of bones as in walking
2. Movement of substances: cardiac muscle pumps blood, smooth muscle helps move food and urine
3. Maintains posture and regulates organ volume: sustained contractions of skeletal muscle maintains sitting or standing position; smooth sphincter muscles control movement of substances through digestive and urinary tracts
4. Heat production: heat is a by-product of skeletal muscle contraction
Type: ES   Page Ref: 225
Topic: FUNCTIONS OF MUSCLE TISSUE

2) Describe and differentiate between a tendon and an aponeurosis.

Answer: A tendon is a band or cord of dense connective tissue that attaches a muscle to the periosteum of a bone. An aponeurosis acts as a tendon, but is a broad, sheetlike layer of dense connective tissue that may attach a muscle to skin, to another muscle, or to the periosteum of a bone.
Type: ES   Page Ref: 226
Topic: SKELETAL MUSCLE TISSUE

3) Describe the events that occur in a sarcomere according to the filament theory of muscle contraction.

Answer: The answer should include the formation of cross bridges, the roles of ATP and calcium, the role of troponin, and the movement of the actin filaments, as on pp. 235.
Type: ES   Page Ref: 235
Topic: SKELETAL MUSCLE: Contraction

4) Compare the microscopic anatomy and the resistance to fatigue of the three types of skeletal muscle fibers.

Answer: Table 9.2, pp. 237
Type: ES   Page Ref: 236
Topic: TYPES OF SKELETAL MUSCLE FIBERS

SHORT ANSWER. Write the word or phrase that best completes each statement or answers the question.

1) Autorhythmicity is the built-in rhythm of smooth muscle and _____ muscle tissue.

Answer: cardiac
Type: SA   Page Ref: 225
Topic: TYPES OF MUSCLE TISSUE

2) Striations are visible when skeletal and _____ muscle tissue are examined microscopically.

Answer: cardiac
Type: SA   Page Ref: 225
Topic: TYPES OF MUSCLE TISSUE

3) The characteristic of muscle tissue that allows it to be stretched without damage to the tissue is _____.

Answer: extensibility
Type: SA   Page Ref: 226
Topic: CHARACTERISTICS OF MUSCLE TISSUE

4) _____ fascia is a subcutaneous layer of loose connective tissue that stores water and fat, and carries nerves and blood vessels to and from skeletal muscles.

Answer: superficial
Type: SA   Page Ref: 226
Topic: SKELETAL MUSCLE TISSUE

5) The connective tissue that surrounds and separates individual skeletal muscle fibers is called _____.

Answer: endomysium
Type: SA   Page Ref: 226
Topic: SKELETAL MUSCLE TISSUE

6) The specialized region of sarcolemma that is located at a neuromuscular junction is called the _____.

Answer: motor end plate
Type: SA   Page Ref: 228
Topic: SKELETAL MUSCLE TISSUE

7) The neurotransmitter released at a neuromuscular junction is _____.

Answer: acetylcholine
Type: SA   Page Ref: 228
Topic: SKELETAL MUSCLE TISSUE

8) A sustained minimal contraction in a skeletal muscle that does not produce movement is muscle _____.

Answer: tone
Type: SA   Page Ref: 235
Topic: SKELETAL MUSCLE: Contraction

9) _____ discs in cardiac muscle contain both anchoring and communicating junctions.

Answer: intercalated
Type: SA   Page Ref: 240
Topic: CARDIAC MUSCLE TISSUE

10) _____ muscle cells do not contain sarcomeres.

Answer: smooth
Type: SA   Page Ref: 240
Topic: SMOOTH MUSCLE TISSUE

11) The muscle tissue with the best ability to regenerate is _____ muscle.

Answer: smooth
Type: SA   Page Ref: 240
Topic: REGENERATION

MATCHING. Choose the item in Column 2 that best matches each item in Column 1.

Match the types of muscle tissue in Column 1 with their characteristics in Column 2.
1) Column 1: skeletal
Column 2: voluntary, striated

Answer: voluntary, striated
Type: MA   Page Ref: 242
Topic: TYPES OF MUSCLE TISSUE

2)   Column 1: cardiac
      Column 2: involuntary, striated
      Foil: voluntary, nonstriated

      Answer: involuntary, striated
      Type: MA   Page Ref: 242
      Topic: TYPES OF MUSCLE TISSUE

3)   Column 1: smooth
      Column 2: involuntary, nonstriated

      Answer: involuntary, nonstriated
      Type: MA   Page Ref: 242
      Topic: TYPES OF MUSCLE TISSUE

4)   Column 1: skeletal muscle
      Column 2: multinucleate

      Answer: multinucleate
      Type: MA   Page Ref: 242
      Topic: TYPES OF MUSCLE TISSUE

5)   Column 1: smooth muscle
      Column 2: no transverse tubules
      Foil: no muscle tone

      Answer: no transverse tubules
      Type: MA   Page Ref: 242
      Topic: TYPES OF MUSCLE TISSUE

6)   Column 1: cardiac muscle
      Column 2: branched cells

      Answer: branched cells
      Type: MA   Page Ref: 242
      Topic: TYPES OF MUSCLE TISSUE

Match the features of a neuromuscular junction (NMJ) in Column 1 with the descriptions in Column 2.

7)   Column 1: axon terminal
      Column 2: bulb-shaped ending of a
                   motor neuron axon

      Answer: bulb-shaped ending of a motor neuron axon
      Type: MA   Page Ref: 228
      Topic: SKELETAL MUSCLE TISSUE

8)   Column 1: synaptic vesicle
        Column 2: sac that stores acetylcholine
        Foil: a small space that
                   separates the cell membrane
                   of a motor neuron from the
                   sarcolemma of a skeletal
                   muscle cell

Answer: sac that stores acetylcholine
Type: MA   Page Ref: 228
Topic: SKELETAL MUSCLE TISSUE

9)   Column 1: motor end plate
        Column 2: region of sarcolemma at the
                   NMJ
        Foil: portion of axon terminal
                   membrane that releases
                   acetylcholine into the cleft

Answer: region of sarcolemma at the NMJ
Type: MA   Page Ref: 228
Topic: SKELETAL MUSCLE TISSUE

10)   Column 1: acetylcholine receptors
        Column 2: integral proteins of motor
                   end plate; bind with
                   acetylcholine
        Foil: integral proteins of axon
                   membrane; bind with
                   acetylcholine

Answer: integral proteins of motor end plate; bind with acetylcholine
Type: MA   Page Ref: 228
Topic: SKELETAL MUSCLE TISSUE

Match the types of skeletal muscle fibers in Column 1 with their descriptions in Column 2.

11)   Column 1: slow fibers
        Column 2: smallest in diameter

Answer: smallest in diameter
Type: MA   Page Ref: 237
Topic: TYPES OF SKELETAL MUSCLE FIBERS

12)   Column 1: fast fibers
        Column 2: largest in diameter

Answer: largest in diameter
Type: MA   Page Ref: 237
Topic: TYPES OF SKELETAL MUSCLE FIBERS

13)　Column 1: fast fibers
　　　Column 2: adapted for intense
　　　　　　　movements of short
　　　　　　　duration, such as weight
　　　　　　　lifting

　　　Answer: adapted for intense movements of short duration, such as weight lifting
　　　Type: MA　　Page Ref: 237
　　　Topic: TYPES OF SKELETAL MUSCLE FIBERS

Match the types of skeletal muscle fibers in Column 1 with their descriptions in Column 2.
14)　Column 1: slow fibers
　　　Column 2: adapted for maintaining
　　　　　　　posture and endurance-type
　　　　　　　activities

　　　Answer: adapted for maintaining posture and endurance-type activities
　　　Type: MA　　Page Ref: 237
　　　Topic: TYPES OF SKELETAL MUSCLE FIBERS

15)　Column 1: intermediate fibers
　　　Column 2: fatigue resistant; adapted
　　　　　　　for activities such as
　　　　　　　walking and sprinting

　　　Answer: fatigue resistant; adapted for activities such as walking and sprinting
　　　Type: MA　　Page Ref: 237
　　　Topic: TYPES OF SKELETAL MUSCLE FIBERS

16)　Column 1: fast fibers
　　　Column 2: low myoglobin content,
　　　　　　　relatively few capillaries

　　　Answer: low myoglobin content, relatively few capillaries
　　　Type: MA　　Page Ref: 237
　　　Topic: TYPES OF SKELETAL MUSCLE FIBERS

CHAPTER 10   The Muscular System

MULTIPLE CHOICE. Choose the one alternative that best completes the statement or answers the question.

1)   The insertion of a muscle:
1. in a limb is usually proximal to the origin.
2. does not move when the muscle contracts.
3. is the movable point of attachment of a muscle.
4. attaches a muscle to a bone or skin.
A) 1, 2, 3, 4   B) 1, 2, 4   C) 1, 3   D) 3, 4

Answer: D
Type: MC   Page Ref: 249
Topic: ORIGIN AND INSERTION

2)   The strength of movement produced by a muscle depends upon how close to the joint it is attached. A muscle attached farther away will produce a more powerful movement than one attached nearer the joint.
A) Both statements are true.
B) Both statements are false.
C) Statement 1 is true; statement 2 is false.
D) Statement 2 is true; statement 1 is false.

Answer: A
Type: MC   Page Ref: 249
Topic: LEVERS

3)   Most skeletal muscles produce movement as a component of a _____–class lever.
A) first   B) second   C) third   D) fourth

Answer: C
Type: BI   Page Ref: 252
Topic: LEVERS

4)   The serratus anterior muscle is named according to:
A) the direction of fibers.   B) shape and size.
C) size and location.   D) shape and location.

Answer: D
Type: MC   Page Ref: 254
Topic: NAMING SKELETAL MUSCLES

5)   Pennate muscles:
A) have a large number of fascicles attached to the tendon.
B) have short muscle fibers attached to a tendon that is almost as long as the entire muscle.
C) have a relatively small range of movement but a greater power than parallel or fusiform muscles.
D) all of the above.

Answer: D
Type: MC   Page Ref: 252
Topic: ARRANGEMENT OF FASCICLES

6)   Which of the following is/are true?
        1. Synergists are muscles that oppose the agonist.
        2. Agonists contract when antagonists relax.
        3. The quadriceps femoris and the hamstrings are antagonists.
        A) 1, 2, 3   B) 2, 3   C) 1, 2   D) 2 only

        Answer: B
        Type: MC   Page Ref: 252
        Topic: GROUP ACTIONS

7)   Which of the following pairs of terms is most closely matched?
        A) orbicularis oculi; shapes lips for speech
        B) corrugator supercilli; surprise
        C) platysma; pouting
        D) zygomaticus major; frown

        Answer: C
        Type: MC   Page Ref: 260
        Topic: MUSCLES OF FACIAL EXPRESSION

8)   The extrinsic muscles of the eyeball are innervated by:
        A) oculomotor (III) nerves.   B) abducens (VI) nerves.
        C) trochlear (IV) nerves.   D) all of the above.

        Answer: D
        Type: MC   Page Ref: 264
        Topic: MUSCLES THAT MOVE THE EYEBALLS

9)   Two muscles that insert on the angle and ramus of the mandible, and that elevate
        and assist in side-to-side movements of the mandible are:
        1. masseter
        2. medial pterygoid
        3. lateral pterygoid
        4. temporalis
        A) 1, 2   B) 2, 3   C) 3, 4   D) 1, 4

        Answer: A
        Type: MC   Page Ref: 267
        Topic: MUSCLES THAT MOVE THE MANDIBLE

10)   The diagastric, stylohyoid, mylohyoid, and geniohyoid muscles have which of the
        following in common?
        A) insertion on the body of the hyoid bone
        B) origin on the body of the hyoid bone
        C) depress the hyoid bone
        D) elevate the tongue

        Answer: A
        Type: MC   Page Ref: 272
        Topic: MUSCLES OF THE FLOOR OF THE ORAL CAVITY

11) The _____ muscle extends from the sternum to the thyroid cartilage and acts to depress the larynx.
A) sternothyroid   B) sternohyoid   C) omohyoid   D) thyrohyoid

Answer: A
Type: BI   Page Ref: 274
Topic: MUSCLES OF THE LARYNX

12) The following muscles all have attachments to ribs or their costal cartilages. Which of them are *not* used to produce the movements of normal breathing?
1. external intercostals
2. internal intercostals
3. external obliques
4. internal obliques
5. diaphragm
A) 2, 3, 4   B) 3, 4   C) 1, 3, 5   D) 2, 4

Answer: A
Type: MC   Page Ref: 285
Topic: MUSCLES USED IN VENTILATION

13) The central portion of the diaphragm is an aponeurosis that serves as the tendon of insertion for the diaphragm. The tendon is called:
A) linea alba.   B) central tendon.
C) intermediate tendon.   D) rectus sheath.

Answer: B
Type: MC   Page Ref: 285
Topic: MUSCLES USED IN VENTILATION

14) Which of the muscles of the pelvic floor constricts the anus, urethra, and vagina?
A) pubococcygeus   B) iliococcygeus   C) coccygeus   D) both A and B

Answer: D
Type: MC   Page Ref: 287
Topic: MUSCLES OF THE PELVIC FLOOR

15) The perineum consists of an anterior _____ triangle and a posterior _____ triangle.
A) anal; urogenital   B) urinary; genital   C) urogenital; anal   D) genital; urinary

Answer: C
Type: BI   Page Ref: 289
Topic: MUSCLES OF THE PERINEUM

16) The pectoralis minor muscle:
A) inserts on the greater tubercle of the humerus.
B) inserts on the acromion of the scapula.
C) originates from ribs 3 to 5.
D) originates from the clavicle and sternum.

Answer: C
Type: MC   Page Ref: 291
Topic: MUSCLES THAT MOVE THE PECTORAL GIRDLE

17) The following muscles all insert on the scapula. Which one does *not* adduct the scapula?
   A) rhomboideus major
   B) rhomboideus minor
   C) trapezius
   D) serratus anterior

   Answer: D
   Type: MC   Page Ref: 291
   Topic: MUSCLES THAT MOVE THE PECTORAL GIRDLE

18) Which of the following pairs of terms is most closely matched?
   A) subclavius; subclavian nerve
   B) rhomboideus major; dorsal scapular nerve
   C) serratus anterior; long thoracic nerve
   D) all of the above

   Answer: D
   Type: MC   Page Ref: 291
   Topic: MUSCLES THAT MOVE THE PECTORAL GIRDLE

19) The greater tubercle of the humerus is the point of insertion of which of the following muscles?
   A) infraspinatus   B) deltoid   C) subscapularis   D) coracobrachialis

   Answer: A
   Type: MC   Page Ref: 295
   Topic: MUSCLES THAT MOVE THE HUMERUS

20) Of the muscles that move the forearm, two have points of origin on the scapula. They are the biceps brachii and the:
   A) brachialis.   B) brachioradialis.   C) triceps brachii.   D) anconeus.

   Answer: C
   Type: MC   Page Ref: 300
   Topic: MUSCLES THAT MOVE THE RADIUS AND ULNA

21) Which muscle originates from the medial epicondyle of the humerus and turns the forearm so that the palm faces posteriorly?
   A) pronator quadratus
   B) pronator teres
   C) anconeus
   D) supinator

   Answer: B
   Type: MC   Page Ref: 300
   Topic: MUSCLES THAT MOVE THE RADIUS AND ULNA

22) Whereas superficial flexors in the anterior compartment of the forearm originate from the _____ epicondyle of the humerus, the superficial extensors in the posterior compartment of the forearm originate from the _____ epicondyle of the humerus.
   A) lateral; medial   B) lateral; lateral   C) medial; lateral   D) medial; medial

   Answer: C
   Type: BI   Page Ref: 306, 307
   Topic: MUSCLES THAT MOVE THE WRIST, HAND, AND DIGITS

23) The extensor digitorum extends the phalanges. The extensor digiti minimi extends the distal phalanges only.
A) Both statements are true.
B) Both statements are false.
C) The first statement is true; the second is false.
D) The second statement is true; the first is false.

Answer: C
Type: MC   Page Ref: 307
Topic: MUSCLES THAT MOVE THE WRIST, HAND, AND DIGITS

24) The erector spinae is a group of muscles of the back, including which three of the following groups?
1. iliocostalis
2. spinalis
3. segmental
4. longissimus
5. scalene
A) 1, 2, 3   B) 1, 2, 4   C) 2, 3, 5   D) 1, 4, 5

Answer: B
Type: MC   Page Ref: 316
Topic: MUSCLES THAT MOVE THE VERTEBRAL COLUMN

25) The psoas major and iliacus muscles have a common insertion on the femur. These two muscles are involved in _____ of the thigh.
A) flexion   B) extension   C) adduction   D) abduction

Answer: A
Type: BI   Page Ref: 322
Topic: MUSCLES THAT MOVE THE FEMUR

26) The tensor fasciae latae originates from the _____ and inserts on the _____.
A) ilium, femur   B) ilium, tibia   C) ischium, femur   D) ischium, tibia

Answer: B
Type: BI   Page Ref: 323
Topic: MUSCLES THAT MOVE THE FEMUR

27) Which of the following statements regarding the quadriceps femoris is correct?
A) Rectus femoris originates from the anterior inferior iliac spine.
B) Vastus lateralis inserts on the lateral condyle of the tibia.
C) All four muscles flex the thigh.
D) Vastus medialis originates from the lesser trochanter.

Answer: A
Type: MC   Page Ref: 331
Topic: MUSCLES THAT ACT ON THE FEMUR, TIBIA, AND FIBULA

28) Which of the following pairs of terms is most closely matched?
A) gracilis; obturator nerve   B) hamstrings; femoral nerve
C) quadriceps femoris; sciatic nerve   D) all of the above

Answer: A
Type: MC   Page Ref: 330
Topic: MUSCLES THAT ACT ON THE FEMUR, TIBIA, AND FIBULA

29) The muscles of the posterior compartment of the thigh _____ the thigh and _____ the leg.
   A) flex, extend   B) extend, flex   C) adduct, flex   D) abduct, extend

   Answer: B
   Type: BI   Page Ref: 330
   Topic: MUSCLES THAT ACT ON THE FEMUR, TIBIA, AND FIBULA

30) Choose the true statement(s):
   A) All three hamstring muscles have a site of origin on the ischial tuberosity.
   B) All three hamstring muscles are innervated by the tibial branch of the sciatic nerve.
   C) The biceps femoris is the lateral hamstring muscle.
   D) All of the above are true.

   Answer: D
   Type: MC   Page Ref: 331
   Topic: MUSCLES THAT ACT ON THE FEMUR, TIBIA, AND FIBULA

31) Muscles that insert via the calcaneal tendon do *not* include:
   A) plantaris.   B) gastrocnemius.
   C) tibialis posterior.   D) soleus.

   Answer: C
   Type: MC   Page Ref: 333
   Topic: MUSCLES THAT MOVE THE FOOT AND TOES

TRUE/FALSE. Write 'T' if the statement is true and 'F' if the statement is false.

1) The most common type of lever in the body is the first–class lever designated as EFR.

   Answer: FALSE
   Type: TF   Page Ref: 251
   Topic: LEVERS

2) Muscles with either parallel or fusiform arrangements of fasciculi tend to have longer fibers than do pennate muscles.

   Answer: TRUE
   Type: TF   Page Ref: 252
   Topic: ARRANGEMENT OF FASCICLES

3) The range of movement at a joint depends upon the point of attachment of the muscle and the length of its fibers.

   Answer: TRUE
   Type: TF   Page Ref: 249
   Topic: LEVERS

4) The masseter produces chewing movements of the mandible.

   Answer: TRUE
   Type: TF   Page Ref: 267
   Topic: MUSCLES THAT MOVE THE MANDIBLE

5)   The lateral rectus is an extrinsic muscle of the tongue.
     Answer: FALSE
     Type: TF   Page Ref: 270
     Topic: MUSCLES THAT MOVE THE TONGUE

6)   The inferior constrictor, middle constrictor, and superior constrictor are muscles that move the wall of the pharynx.
     Answer: TRUE
     Type: TF   Page Ref: 277
     Topic: MUSCLES OF THE PHARYNX

7)   The sternocleidomastoid muscle inserts on the temporal bone.
     Answer: TRUE
     Type: TF   Page Ref: 280
     Topic: MUSCLES THAT MOVE THE HEAD

8)   The rectus sheath and linea alba are formed by the aponeuroses of the external obliques, internal obliques, and transversus abdominis muscles.
     Answer: TRUE
     Type: TF   Page Ref: 281
     Topic: MUSCLES THAT ACT ON THE ABDOMINAL WALL

9)   The ischiocavernosus muscles maintain erection of the penis and erection of the clitoris.
     Answer: TRUE
     Type: TF   Page Ref: 289
     Topic: MUSCLES OF THE PERINEUM

10) The serratus anterior muscle inserts on the lateral border of the scapula.
     Answer: FALSE
     Type: TF   Page Ref: 291
     Topic: MUSCLES THAT MOVE THE PECTORAL GIRDLE

11) The pectoralis minor muscle and a portion of the trapezius depress the scapula.
     Answer: TRUE
     Type: TF   Page Ref: 291
     Topic: MUSCLES THAT MOVE THE PECTORAL GIRDLE

12) The musculocutaneous nerve innervates the biceps brachii and brachialis muscles.
     Answer: TRUE
     Type: TF   Page Ref: 300
     Topic: MUSCLES THAT MOVE THE RADIUS AND ULNA

13) Opposition is a movement of the human thumb that is important for the creation and utilization of tools.
     Answer: TRUE
     Type: TF   Page Ref: 313
     Topic: INSTRINSIC MUSCLES OF THE HAND

14) The scalenes originate on the first and second ribs and insert on the third through sixth thoracic vertebrae.

Answer: FALSE
Type: TF   Page Ref: 316
Topic: MUSCLES THAT MOVE THE VERTEBRAL COLUMN

15) The three gluteus muscles act as a group to adduct the thigh.

Answer: FALSE
Type: TF   Page Ref: 322
Topic: MUSCLES THAT MOVE THE FEMUR

16) The piriformis, obturator internus, and obturator externus insert on or near the greater trochanter, and all three laterally rotate the thigh.

Answer: TRUE
Type: TF   Page Ref: 322
Topic: MUSCLES THAT MOVE THE FEMUR

17) Tibialis anterior and tibialis posterior muscles both invert the foot.

Answer: TRUE
Type: TF   Page Ref: 334
Topic: MUSCLES THAT MOVE THE FOOT AND TOES

ESSAY. Write your answer in the space provided or on a separate sheet of paper.

1) Using the terms prime mover, antagonist, synergist, and fixator, describe the various roles muscles may play in a group.

Answer: Definitions and examples are given on pp. 252.
Type: ES   Page Ref: 252
Topic: GROUP ACTIONS

2) List the muscles that act on the mandible to open the mouth and those that act on the mandible to close the mouth. Which are more forceful? Why?

Answer: Masseter, temporalis, and medial pterygoid close the mouth. Lateral pterygoid, digastric, mylohyoid, and geniohyoid open the mouth. Closing the mouth is more forceful. Masseter and temporalis are large, powerful muscles whose main action is to elevate the mandible. All muscles that act to depress the mandible have other primary actions.
Type: ES   Page Ref: 267
Topic: MUSCLES THAT MOVE THE MANDIBLE

3) Name three muscles that insert on the mastoid process. All three muscles move the head. Which one of these is antagonistic to the other two in the action of rotating the head?

Answer: sternocleidomastoid, splenius capitis, longissimus capitis. The sternocleidomastoid is the antagonist. It rotates the face away from the contracting muscle, whereas the other two cause a rotation toward the same side as the contracting muscle.
Type: ES   Page Ref: 280
Topic: MUSCLES THAT MOVE THE HEAD

4)   Name and describe the location of each of the muscles of the three compartments of the thigh: anterior, medial, and posterior.

Answer: see Exhibit 10.20.
Type: ES   Page Ref: 330
Topic: MUSCLES THAT ACT ON THE FEMUR, TIBIA, AND FIBULA

SHORT ANSWER.   Write the word or phrase that best completes each statement or answers the question.

1)   The portion of a muscle between the origin and the insertion is called the _____ of the muscle.

Answer: belly
Type: SA   Page Ref: 249
Topic: ORIGIN AND INSERTION

2)   To produce movement, a bone acts as a lever and the joint acts as the _____.

Answer: fulcrum
Type: SA   Page Ref: 249
Topic: LEVERS

3)   A circular muscle that encloses an orifice is a/an _____ muscle.

Answer: sphincter
Type: SA   Page Ref: 252
Topic: ARRANGEMENT OF FASCICLES

4)   Prime mover is to _____ as flexor is to extensor.

Answer: antagonist
Type: SA   Page Ref: 252
Topic: GROUP ACTIONS

5)   The tendon that unites the frontalis and occipitalis muscles is called the _____.

Answer: galea aponeurotica
Type: SA   Page Ref: 259
Topic: MUSCLES OF FACIAL EXPRESSION

6)   The levator palpebrae superioris muscle elevates the _____.

Answer: upper eyelid
Type: SA   Page Ref: 260
Topic: MUSCLES OF FACIAL EXPRESSION

7)   The muscle that acts on your left eyeball to move it as you read this line is the _____.

Answer: medial rectus
Type: SA   Page Ref: 265
Topic: MUSCLES THAT MOVE THE EYEBALLS

8)   All muscles that move the mandible are innervated by the cranial nerve called the
      _____.
      Answer: trigeminal (V)
      Type: SA   Page Ref: 267
      Topic: MUSCLES THAT MOVE THE MANDIBLE

9)   The genioglossus muscle could be considered a socially unacceptable muscle, since its
      action is to _____ the tongue.
      Answer: protract
      Type: SA   Page Ref: 270
      Topic: MUSCLES THAT MOVE THE TONGUE

10)   The four muscles that form the anterolateral abdominal wall have one action in
      common; they all _____.
      Answer: compress the abdomen
      Type: SA   Page Ref: 281
      Topic: MUSCLES THAT ACT ON THE ABDOMINAL WALL

11)   Branches of the _____ nerve are involved in the innervation of all perineal muscles.
      Answer: pudendal
      Type: SA   Page Ref: 289
      Topic: MUSCLES OF THE PERINEUM

12)   The rotator cuff muscles are _____, _____, _____, and _____.
      Answer: supraspinatus, infraspinatus, subscapularis, teres minor
      Type: SA   Page Ref: 295
      Topic: MUSCLES THAT MOVE THE HUMERUS

13)   Muscles that extend the forearm are located on the _____ surface of the _____.
      Answer: posterior, humerus
      Type: SA   Page Ref: 300
      Topic: MUSCLES THAT MOVE THE RADIUS AND ULNA

14)   The flexor carpi _____ flexes and adducts the wrist.
      Answer: ulnaris
      Type: SA   Page Ref: 306
      Topic: MUSCLES THAT MOVE THE WRIST, HAND, AND DIGITS

15)   The four intrinsic muscles of the hand that produce thumb movements are, as a
      group, called _____ muscles.
      Answer: thenar
      Type: SA   Page Ref: 312
      Topic: INSTRINSIC MUSCLES OF THE HAND

16)   The muscle used for extending the thigh, as in climbing stairs, is the _____.
      Answer: gluteus maximus
      Type: SA   Page Ref: 322
      Topic: MUSCLES THAT MOVE THE FEMUR

MATCHING. Choose the item in Column 2 that best matches each item in Column 1.

Match the muscles in Column 1 with their actions in Column 2.

1) Column 1: flexor hallucis longus
   Column 2: flexes great toe
   Foil: flexes thumb

   Answer: flexes great toe
   Type: MA   Page Ref: 334
   Topic: MUSCLES THAT MOVE THE FOOT AND TOES

2) Column 1: tibialis anterior
   Column 2: dorsiflexes foot

   Answer: dorsiflexes foot
   Type: MA   Page Ref: 334
   Topic: MUSCLES THAT MOVE THE FOOT AND TOES

3) Column 1: gastrocnemius
   Column 2: flexes leg

   Answer: flexes leg
   Type: MA   Page Ref: 335
   Topic: MUSCLES THAT MOVE THE FOOT AND TOES

4) Column 1: rectus femoris
   Column 2: extends leg

   Answer: extends leg
   Type: MA   Page Ref: 331
   Topic: MUSCLES THAT MOVE THE FEMUR, TIBIA, AND FIBULA

5) Column 1: flexor digitorum longus
   Column 2: flexes toes

   Answer: flexes toes
   Type: MA   Page Ref: 335
   Topic: MUSCLES THAT MOVE THE FOOT AND TOES

6) Column 1: plantar interossei
   Column 2: adduct toes

   Answer: adduct toes
   Type: MA   Page Ref: 341
   Topic: INTRINSIC MUSCLES OF THE FOOT

7) Column 1: dorsal interossei
   Column 2: abducts toes

   Answer: abducts toes
   Type: MA   Page Ref: 341
   Topic: INTRINSIC MUSCLES OF THE FOOT

CHAPTER 11   Surface Anatomy

MULTIPLE CHOICE. Choose the one alternative that best completes the statement or answers the question.

1)   Tragus, helix, lobule, and concha are features of the:
      A) ear.   B) nose.   C) eye.   D) neck.

      Answer: A
      Type: MC   Page Ref: 354
      Topic: SURFACE ANATOMY OF THE HEAD

2)   Which of the following statements is *not* true?
      A) The upper and lower palpebrae are lined by conjunctiva.
      B) The lacrimal caruncle is located at the lateral union of the upper and lower
         palpebrae.
      C) Eyelashes consist of hairs that are usually arranged in two or three rows.
      D) Sclera is the "white" of the eye.

      Answer: B
      Type: MC   Page Ref: 354
      Topic: SURFACE ANATOMY OF THE HEAD

3)   In the neck, the external jugular vein can be seen in the:
      A) midline.   B) posterior triangle.
      C) anterior triangle.   D) submandibular triangle.

      Answer: B
      Type: MC   Page Ref: 357
      Topic: SURFACE ANATOMY OF THE NECK

4)   The triangle of auscultation is bordered by two muscles of the back:
      A) latissimus dorsi and teres major.   B) erector spinae and teres major.
      C) trapezius and infraspinatus.   D) trapezius and latissimus dorsi.

      Answer: D
      Type: MC   Page Ref: 358
      Topic: SURFACE ANATOMY OF THE TRUNK

5)   The sternal angle, a surface landmark of the chest:
      A) lies at the junction of the clavicles and manubrium.
      B) can be used to locate the costal cartilages of the second ribs.
      C) is a depression on the superior border of the manubrium.
      D) is found at the inferior end of the sternum.

      Answer: B
      Type: MC   Page Ref: 359
      Topic: SURFACE ANATOMY OF THE TRUNK

6)   The arch of the aorta is posterior to the:
      A) manubrium of sternum.   B) sternal angle.
      C) body of sternum.   D) costal margin.

      Answer: A
      Type: MC   Page Ref: 359
      Topic: SURFACE ANATOMY OF THE TRUNK

7) McBurney's point, a landmark used in locating the appendix:
A) is inferior and lateral to the umbilicus.
B) is superior and medial to the right anterior superior iliac spine.
C) is closer to the right anterior superior iliac spine than it is to the umbilicus.
D) all of the above.

Answer: D
Type: MC   Page Ref: 362
Topic: SURFACE ANATOMY OF THE TRUNK

8) The cubital fossa:
A) contains the ulnar nerve.   B) contains the axillary artery.
C) is posterior to the olecranon.   D) none of the above.

Answer: D
Type: MC   Page Ref: 365
Topic: SURFACE ANATOMY OF THE UPPER LIMB

9) The popliteal fossa is bordered by which of the following muscles?
1. biceps femoris
2. semimembranosus
3. adductor magnus
4. gastrocnemius
5. semitendinosus
A) 1, 2, 3, 4, 5   B) 1, 2, 4, 5   C) 1, 3, 4, 5   D) 2, 4, 5

Answer: B
Type: MC   Page Ref: 369
Topic: SURFACE ANATOMY OF THE LOWER LIMB

TRUE/FALSE.   Write 'T' if the statement is true and 'F' if the statement is false.

1) The oral region of the face includes the eyeballs, eyebrows, and eyelids.
Answer: FALSE
Type: TF   Page Ref: 350
Topic: SURFACE ANATOMY OF THE HEAD

2) The trachea can be palpated in the suprasternal notch of the sternum.
Answer: TRUE
Type: TF   Page Ref: 359
Topic: SURFACE ANATOMY OF THE TRUNK

3) In the abdomen, the inferior vena cava is located to the right of the abdominal aorta.
Answer: TRUE
Type: TF   Page Ref: 362
Topic: SURFACE ANATOMY OF THE TRUNK

4) The scapulae are on the back, at the level of ribs 2 through 7.
Answer: TRUE
Type: TF   Page Ref: 358
Topic: SURFACE ANATOMY OF THE TRUNK

5) The deltoid muscle forms the anterior axillary fold.

Answer: FALSE
Type: TF   Page Ref: 363
Topic: SURFACE ANATOMY OF THE UPPER LIMB

6) The ulnar nerve passes through a groove behind the lateral epicondyle of the humerus.

Answer: FALSE
Type: TF   Page Ref: 365
Topic: SURFACE ANATOMY OF THE UPPER LIMB

7) In the anatomical position, the thenar eminence of the palm is medial to the hypothenar eminence.

Answer: FALSE
Type: TF   Page Ref: 368
Topic: SURFACE ANATOMY OF THE UPPER LIMB

8) The depression between the two buttocks is the gluteal fold.

Answer: FALSE
Type: TF   Page Ref: 368
Topic: SURFACE ANATOMY OF THE LOWER LIMB

9) The inguinal ligament forms the superior border of the femoral triangle.

Answer: TRUE
Type: TF   Page Ref: 369
Topic: SURFACE ANATOMY OF THE LOWER LIMB

ESSAY. Write your answer in the space provided or on a separate sheet of paper.

1) Describe the boundaries of the anterior and posterior triangles of the neck.

Answer: The borders of the anterior triangle are as follows:
        superior: mandible
        inferior: sternum
        medial: cervical midline
        lateral: anterior border of the sternocleidomastoid muscle
        The borders of the posterior triangle are as follows:
        inferior: clavicle
        anterior: posterior border of the sternocleidomastoid muscle
        posterior: anterior border of the trapezius muscle
Type: ES   Page Ref: 357, 358
Topic: SURFACE ANATOMY OF THE NECK

2) Name and locate the four points (surface markings) that can be used to determine the location and size of the heart.

Answer: 1. inferior left point (apex): at the 5th intercostal space, 9 cm to the left of the midline

2. inferior right point: at the lower border of the costal cartilage of the right 6th rib, about 3 cm to the right of the midline

3. superior right point: at the upper border of the costal cartilage of the right 3rd rib, about 3 cm right of the midline

4. superior left point: at the lower border of the costal cartilage of the left 2nd rib, about 3 cm to the left of the midline

Type: ES   Page Ref: 360
Topic: SURFACE ANATOMY OF THE TRUNK

3) Describe the locations of the muscles of the thigh that are visible as superficial landmarks.

Answer: The locations of the sartorius, quadriceps femoris, adductor longus, and hamstring muscles are described on pp. 369

Type: ES   Page Ref: 369
Topic: SURFACE ANATOMY OF THE LOWER LIMB

SHORT ANSWER. Write the word or phrase that best completes each statement or answers the question.

1) The _____ of the mandible can be palpated in the buccal region.

Answer: ramus
Type: SA   Page Ref: 351
Topic: SURFACE ANATOMY OF THE HEAD

2) The region of the nose between apex and root is the _____.

Answer: dorsum nasi
Type: SA   Page Ref: 355
Topic: SURFACE ANATOMY OF THE HEAD

3) The posterior axillary fold is formed by two muscles, teres major and _____.

Answer: latissimus dorsi
Type: SA   Page Ref: 358
Topic: SURFACE ANATOMY OF THE TRUNK

4) The _____ lumbar vertebra is a landmark for performing a lumbar puncture.

Answer: fourth
Type: SA   Page Ref: 358
Topic: SURFACE ANATOMY OF THE TRUNK

5) The anterior axillary fold is formed by the lateral border of the _____ muscle.

Answer: pectoralis major
Type: SA   Page Ref: 360
Topic: SURFACE ANATOMY OF THE TRUNK

6) The _____ ligament is the lower border of the external oblique muscle.

Answer: inguinal
Type: SA   Page Ref: 369
Topic: SURFACE ANATOMY OF THE TRUNK

7) A vein found at the elbow, used for venipuncture (blood withdrawal for analysis) and transfusions is the _____ vein.

Answer: median cubital
Type: SA   Page Ref: 365
Topic: SURFACE ANATOMY OF THE UPPER LIMB

8) The _____ muscle forms the bulk of the anterior surface of the arm.

Answer: biceps brachii
Type: SA   Page Ref: 364
Topic: SURFACE ANATOMY OF THE UPPER LIMB

9) The tendon of the _____ muscle is visible when the wrist is slightly flexed and the base of the thumb and little finger are drawn together.

Answer: palmaris longus
Type: SA   Page Ref: 365
Topic: SURFACE ANATOMY OF THE UPPER LIMB

10) The three bony prominences found in the elbow region are the _____, the _____, and the _____.

Answer: medial epicondyle, lateral epicondyle, olecranon
Type: SA   Page Ref: 364, 365
Topic: SURFACE ANATOMY OF THE UPPER LIMB

11) When the thumb is bent backward, a depression forms called the _____ between the tendons of the extensor pollicis brevis and extensor pollicis longus.

Answer: anatomical snuffbox
Type: SA   Page Ref: 365
Topic: SURFACE ANATOMY OF THE UPPER LIMB

12) The _____ arch is a venous loop that can be pressed on the posterior surface of the hand.

Answer: dorsal venous
Type: SA   Page Ref: 367
Topic: SURFACE ANATOMY OF THE UPPER LIMB

13) The bony projection on the anterior surface near the proximal end of the leg is the _____.

Answer: tibial tuberosity
Type: SA   Page Ref: 369
Topic: SURFACE ANATOMY OF THE LOWER LIMB

MATCHING. Choose the item in Column 2 that best matches each item in Column 1.

Match the anatomical landmarks and regions in Column 1 with their descriptions or common names in Column 2.

1)   Column 1: suprasternal notch
Column 2: site at which trachea can
be palpated

Answer: site at which trachea can be palpated
Type: MA   Page Ref: 359
Topic: SURFACE ANATOMY OF THE TRUNK

2)   Column 1: thyroid cartilage
Column 2: Adam's apple

Answer: Adam's apple
Type: MA   Page Ref: 355
Topic: SURFACE ANATOMY OF THE NECK

3)   Column 1: posterior triangle of neck
Column 2: anterior border is
sternocleidomastoid

Answer: anterior border is sternocleidomastoid
Type: MA   Page Ref: 358
Topic: SURFACE ANATOMY OF THE NECK

4)   Column 1: trapezius muscle
Column 2: causes a stiff neck when
inflamed

Answer: causes a stiff neck when inflamed
Type: MA   Page Ref: 357
Topic: SURFACE ANATOMY OF THE NECK

5)   Column 1: umbilicus
Column 2: navel

Answer: navel
Type: MA   Page Ref: 361
Topic: SURFACE ANATOMY OF THE TRUNK

6)   Column 1: linea alba
Column 2: extends from xiphoid
process to pubic symphysis

Answer: extends from xiphoid process to pubic symphysis
Type: MA   Page Ref: 362
Topic: SURFACE ANATOMY OF THE TRUNK

7)   Column 1: deltoid muscle
Column 2: site of intramuscular
injection in the shoulder

Answer: site of intramuscular injection in the shoulder
Type: MA   Page Ref: 363
Topic: SURFACE ANATOMY OF THE UPPER LIMB

8)   Column 1: acromion
        Column 2: highest point of shoulder

        Answer: highest point of shoulder
        Type: MA   Page Ref: 363
        Topic: SURFACE ANATOMY OF THE UPPER LIMB

9)   Column 1: cubital fossa
        Column 2: anterior region of elbow

        Answer: anterior region of elbow
        Type: MA   Page Ref: 365
        Topic: SURFACE ANATOMY OF THE UPPER LIMB

10)   Column 1: antebrachium
        Column 2: forearm
        Foil: arm

        Answer: forearm
        Type: MA   Page Ref: 365
        Topic: SURFACE ANATOMY OF THE UPPER LIMB

11)   Column 1: knuckles
        Column 2: distal ends of metacarpals

        Answer: distal ends of metacarpals
        Type: MA   Page Ref: 367
        Topic: SURFACE ANATOMY OF THE UPPER LIMB

12)   Column 1: manus
        Column 2: hand
        Foil: wrist

        Answer: hand
        Type: MA   Page Ref: 367
        Topic: SURFACE ANATOMY OF THE UPPER LIMB

13)   Column 1: posterior superior iliac
                 spine
        Column 2: landmark for the inferior
                 limit of cerebrospinal fluid
                 around the spinal cord
        Foil: bears body weight when
                 sitting

        Answer: landmark for the inferior limit of cerebrospinal fluid around the spinal cord
        Type: MA   Page Ref: 362
        Topic: SURFACE ANATOMY OF THE TRUNK

14)   Column 1: vastus medialis
        Column 2: anteromedial thigh

        Answer: anteromedial thigh
        Type: MA   Page Ref: 369
        Topic: SURFACE ANATOMY OF THE LOWER LIMB

15) Column 1: collum
Column 2: neck

Answer: neck
Type: MA   Page Ref: 355
Topic: SURFACE ANATOMY OF THE NECK

Match the cranial and facial regions in Column 1 with their descriptions in Column 2.

16) Column 1: frontal region
Column 2: forms the front of the skull

Answer: forms the front of the skull
Type: MA   Page Ref: 350
Topic: SURFACE ANATOMY OF THE HEAD

17) Column 1: temporal region
Column 2: forms the side of the skull

Answer: forms the side of the skull
Type: MA   Page Ref: 350
Topic: SURFACE ANATOMY OF THE HEAD

18) Column 1: occipital region
Column 2: forms the base of the skull

Answer: forms the base of the skull
Type: MA   Page Ref: 350
Topic: SURFACE ANATOMY OF THE HEAD

19) Column 1: mental region
Column 2: anterior portion of the
                mandible
Foil: region of the nose

Answer: anterior portion of the mandible
Type: MA   Page Ref: 350
Topic: SURFACE ANATOMY OF THE HEAD

20) Column 1: buccal region
Column 2: region of the cheek
Foil: region of the mouth

Answer: region of the cheek
Type: MA   Page Ref: 350
Topic: SURFACE ANATOMY OF THE HEAD

21) Column 1: infraorbital region
Column 2: the region inferior to the
                orbit
Foil: includes the eyeballs,
                eyebrows, and eyelids

Answer: the region inferior to the orbit
Type: MA   Page Ref: 350
Topic: SURFACE ANATOMY OF THE HEAD

22) Column 1: auricular region
    Column 2: region of the external ear

    Answer: region of the external ear
    Type: MA   Page Ref: 350
    Topic: SURFACE ANATOMY OF THE HEAD

CHAPTER 12   The Cardiovascular System: Blood

MULTIPLE CHOICE. Choose the one alternative that best completes the statement or answers the question.

1)   The components of the cardiovascular system include:
      1. interstitial fluid
      2. blood vessels
      3. heart
      4. lymphatic vessels
      5. blood
      A) 1, 2, 3, 4, 5   B) 2, 3, 4, 5   C) 2, 3, 5   D) 1, 2, 3, 5

      Answer: C
      Type: MC   Page Ref: 377
      Topic: CARDIOVASCULAR SYSTEM: Introduction

2)   Blood protects against:
      1. fluid loss
      2. pH imbalance
      3. UV radiation damage
      4. microbes and toxins
      5. fluctuations in body temperature
      A) 1, 2, 3, 4, 5   B) 1, 3, 4, 5   C) 1, 2, 4, 5   D) 2, 4, 5

      Answer: C
      Type: MC   Page Ref: 377
      Topic: FUNCTIONS OF BLOOD

3)   Which of the following is *not* a physical characteristic of blood?
      A) pH range of 7.25 to 7.55
      B) temperature of 38 degrees Celsius
      C) average volume of 5–6 liters in a male
      D) 0.9% salt concentration

      Answer: A
      Type: MC   Page Ref: 377
      Topic: PHYSICAL CHARACTERISTICS OF BLOOD

4)   Sites of hemopoiesis in the adult include all of the following *except*:
      A) liver.   B) sternum.   C) cranial bones.   D) head of femur.

      Answer: A
      Type: MC   Page Ref: 379
      Topic: FORMATION OF BLOOD CELLS

5) The heme portion of hemoglobin contains iron that binds reversibly with oxygen. In addition to oxygen and carbon dioxide transport, hemoglobin functions in blood pressure regulation by ferrying nitric oxide throughout the body.
A) Both statements are true.
B) Both statements are false.
C) The first statement is true; the second is false.
D) The second statement is true; the first is false.

Answer: A
Type: MC   Page Ref: 380–382
Topic: RED BLOOD CELLS

6) The life span of a red blood cell is approximately:
A) 30 days.   B) two months.   C) three months.   D) four months.

Answer: D
Type: MC   Page Ref: 383
Topic: RED BLOOD CELLS

7) During which stage of RBC formation is the nucleus ejected from the cell?
A) basophilic erythroblast   B) polychromatophilic erythroblast
C) orthochromatophilic erythroblast D) reticulocyte

Answer: C
Type: MC   Page Ref: 383
Topic: RED BLOOD CELLS

8) Which of the following lead(s) to an increase in the rate of RBC production?
A) hypoxia   B) testosterone
C) erythropoietin   D) all of the above

Answer: D
Type: MC   Page Ref: 383
Topic: RED BLOOD CELLS

9) A, B, and Rh(D) are examples of _____ found in RBC membranes.
A) antibodies   B) agglutinogens   C) granules   D) globulins

Answer: B
Type: BI   Page Ref: 383
Topic: RED BLOOD CELLS

10) Monocytes migrate into body tissues to become phagocytic cells called:
A) macrophages.   B) fibroblasts.   C) lymphocytes.   D) polymorphs.

Answer: A
Type: MC   Page Ref: 385
Topic: WHITE BLOOD CELLS

11) Neutrophils provide weapons against bacteria in the form of:
A) lysozymes.   B) defensins.
C) oxidants.   D) all of the above.

Answer: D
Type: MC   Page Ref: 385
Topic: WHITE BLOOD CELLS

12) The white blood cells that typically increase in number in response to allergic conditions or parasitic infections are the _____.
A) neutrophils   B) eosinophils   C) basophils   D) monocytes

Answer: B
Type: BI   Page Ref: 385
Topic: WHITE BLOOD CELLS

13) Platelets:
A) have a longer life span than red blood cells.
B) develop from worn out red blood cells.
C) are more numerous than red blood cells.
D) promote blood clotting and blood vessel repair.

Answer: D
Type: MC   Page Ref: 386
Topic: PLATELETS

TRUE/FALSE. Write 'T' if the statement is true and 'F' if the statement is false.

1) Blood is a connective tissue.

Answer: FALSE
Type: TF   Page Ref: 377
Topic: FUNCTIONS OF BLOOD

2) Blood transports oxygen and nutrients to body cells, carbon dioxide and wastes away from body cells.

Answer: TRUE
Type: TF   Page Ref: 377
Topic: FUNCTIONS OF BLOOD

3) The temperature of blood is slightly less than normal body temperature.

Answer: FALSE
Type: TF   Page Ref: 377
Topic: PHYSICAL CHARACTERISTICS OF BLOOD

4) Blood is heavier, thicker, and less viscous than water.

Answer: FALSE
Type: TF   Page Ref: 377
Topic: PHYSICAL CHARACTERISTICS OF BLOOD

5) During fetal development, blood cells may arise from tissue in the thymus gland, spleen, liver, yolk sac, and bone marrow.

Answer: TRUE
Type: TF   Page Ref: 379
Topic: FORMATION OF BLOOD CELLS

6) Red blood cells have no mitochondria, and therefore do not utilize any of the oxygen they transport.

Answer: TRUE
Type: TF   Page Ref: 380
Topic: RED BLOOD CELLS

7) Synthesis of hemoglobin first occurs in the polychromatophilic erythroblast stage of RBC formation.

Answer: TRUE
Type: TF   Page Ref: 383
Topic: RED BLOOD CELLS

8) Hemoglobin transports oxygen and carbon dioxide.

Answer: TRUE
Type: TF   Page Ref: 382
Topic: RED BLOOD CELLS

9) Neutrophils, eosinophils, and basophils are agranulocytes.

Answer: FALSE
Type: TF   Page Ref: 383
Topic: WHITE BLOOD CELLS

10) T cells respond to foreign invaders by producing antibodies.

Answer: FALSE
Type: TF   Page Ref: 385
Topic: WHITE BLOOD CELLS

11) Metamegakaryocytes are cells that break into 2000–3000 platelets.

Answer: TRUE
Type: TF   Page Ref: 386
Topic: PLATELETS

ESSAY. Write your answer in the space provided or on a separate sheet of paper.

1) Describe the three main functions of blood.

Answer: a) transportation of oxygen, carbon dioxide, nutrients, wastes, heat, and hormones
b) regulation of pH, body temperature, and water content of cells
c) protection against blood loss and disease
Type: ES   Page Ref: 377
Topic: FUNCTIONS OF BLOOD

2) List the components of plasma, stating a function of each, or giving a reason why each is found in the plasma.

Answer: See Table 12.1.
Type: ES   Page Ref: 380
Topic: PLASMA

3) Describe the anatomy of a mature red blood cell.

Answer: a) a flexible biconcave disc
b) lacks nucleus, organelles
c) cytosol contains hemoglobin (33% of cell weight)
d) plasma membrane contains antigen (agglutinogen) molecules
Type: ES   Page Ref: 380
Topic: RED BLOOD CELLS

4) Briefly describe the structure of a hemoglobin molecule and relate that to its functions.

Answer: a) four hemes, containing iron as oxygen-binding sites; also have affinity for NO and SNO
b) four protein globin chains that serve as sites for carbon dioxide binding
c) two forms of the molecule: oxyhemoglobin and carbaminohemoglobin
Type: ES   Page Ref: 380–382
Topic: RED BLOOD CELLS

5) Three characteristics of white blood cells that assist in defense against spread of infection are chemotaxis, emigration, and phagocytosis. Define each of these terms in relation to white blood cell function.

Answer: Each term is defined on pp. 385.
Type: ES   Page Ref: 385
Topic: WHITE BLOOD CELLS

SHORT ANSWER. Write the word or phrase that best completes each statement or answers the question.

1) The study of blood and blood-forming tissues is called _____.

Answer: hematology
Type: SA   Page Ref: 377
Topic: CARDIOVASCULAR SYSTEM: Introduction

2) The molecules that are the most abundant and also the smallest of the plasma proteins are the _____.

Answer: albumins
Type: SA   Page Ref: 380
Topic: PLASMA

3) The process of blood cell formation is called _____.

Answer: hemopoiesis or hematopoiesis
Type: SA   Page Ref: 379
Topic: FORMATION OF BLOOD CELLS

4) Fixed macrophages in the _____ and _____ remove worn out red blood cells and platelets from the bloodstream.

Answer: spleen, liver
Type: SA   Page Ref: 383
Topic: RED BLOOD CELLS

5) Hypoxia stimulates the kidneys to produce a hormone called _____ that causes the rate of production of RBC to increase.

Answer: erythropoietin
Type: SA   Page Ref: 383
Topic: RED BLOOD CELLS

6) B cells develop into _____ cells whose function is to produce _____ to inactivate bacterial poisons.

Answer: plasma, antibodies
Type: SA   Page Ref: 385
Topic: WHITE BLOOD CELLS

7) The white blood cells that are the first to respond to tissue destruction caused by bacteria are the _____.

Answer: neutrophils
Type: SA   Page Ref: 385
Topic: WHITE BLOOD CELLS

8) A white blood cell count below normal is called _____.

Answer: leukopenia
Type: SA   Page Ref: 386
Topic: WHITE BLOOD CELLS

9) T cells that destroy invading microbes are also called _____ T cells.

Answer: cytotoxic or killer
Type: SA   Page Ref: 385
Topic: WHITE BLOOD CELLS

10) Phagocytic white blood cells are attracted by microbial toxins and by kinins from damaged tissue. Such cell movement is called _____.

Answer: chemotaxis
Type: SA   Page Ref: 385
Topic: WHITE BLOOD CELLS

MATCHING. Choose the item in Column 2 that best matches each item in Column 1.

Match the cells in Column 1 with their descriptions in Column 2.
1) Column 1: myeloid stem cells
   Column 2: give rise to erythrocytes,
             platelets, granular
             leukocytes, and monocytes
             only

   Answer: give rise to erythrocytes, platelets, granular leukocytes, and monocytes only
   Type: MA   Page Ref: 379
   Topic: FORMATION OF BLOOD CELLS

2) Column 1: hemopoietic stem cells
   Column 2: give rise to all formed
             elements

   Answer: give rise to all formed elements
   Type: MA   Page Ref: 379
   Topic: FORMATION OF BLOOD CELLS

3)   Column 1: lymphoid stem cells
        Column 2: give rise to lymphocytes

        Answer: give rise to lymphocytes
        Type: MA   Page Ref: 379
        Topic: FORMATION OF BLOOD CELLS

4)   Column 1: hemopoietic stem cells
        Column 2: also called pluripotential
                  stem cells

        Answer: also called pluripotential stem cells
        Type: MA   Page Ref: 379
        Topic: FORMATION OF BLOOD CELLS

Match the components of plasma in Column 1 with their descriptions in Column 2.

5)   Column 1: water
        Column 2: solvent of plasma

        Answer: solvent of plasma
        Type: MA   Page Ref: 380
        Topic: PLASMA

6)   Column 1: albumins
        Column 2: smallest plasma proteins,
                  important in maintaining
                  osmotic pressure

        Answer: smallest plasma proteins, important in maintaining osmotic pressure
        Type: MA   Page Ref: 380
        Topic: PLASMA

7)   Column 1: globulins
        Column 2: important in defense
                  against certain
                  microorganisms

        Answer: important in defense against certain microorganisms
        Type: MA   Page Ref: 380
        Topic: PLASMA

8)   Column 1: fibrinogens
        Column 2: essential for blood clot
                  formation

        Answer: essential for blood clot formation
        Type: MA   Page Ref: 380
        Topic: PLASMA

9)   Column 1: ammonium salts
        Column 2: by-product of metabolism,
                  carried by plasma to
                  excretory organs

        Answer: by-product of metabolism, carried by plasma to excretory organs
        Type: MA   Page Ref: 380
        Topic: PLASMA

10) Column 1: inorganic salts
Column 2: essential minerals

Answer: essential minerals
Type: MA   Page Ref: 380
Topic: PLASMA

11) Column 1: amino acids
Column 2: end product of protein
           digestion, carried by plasma
           to body cells
Foil: end product of carbohydrate
           digestion, carried by plasma
           to body cells

Answer: end product of protein digestion, carried by plasma to body cells
Type: MA   Page Ref: 380
Topic: PLASMA

12) Column 1: carbon dioxide
Column 2: by-product of metabolism,
           carried primarily by plasma
           to lungs for removal

Answer: by-product of metabolism, carried primarily by plasma to lungs for removal
Type: MA   Page Ref: 380
Topic: PLASMA

Match the white blood cells in Column 1 with their descriptions in Column 2.
13) Column 1: monocyte
Column 2: large indented nucleus;
           "foamy" cytoplasm
Foil: no nucleus; disc-shaped cell

Answer: large indented nucleus; "foamy" cytoplasm
Type: MA   Page Ref: 384
Topic: WHITE BLOOD CELLS

14) Column 1: lymphocyte
Column 2: large rounded nucleus; blue
           cytoplasm with no visible
           granules

Answer: large rounded nucleus; blue cytoplasm with no visible granules
Type: MA   Page Ref: 384
Topic: WHITE BLOOD CELLS

15) Column 1: eosinophil
Column 2: bilobed nucleus; large red-
           orange granules in
           cytoplasm

Answer: bilobed nucleus; large red-orange granules in cytoplasm
Type: MA   Page Ref: 383
Topic: WHITE BLOOD CELLS

16) Column 1: basophil
   Column 2: bilobed or S-shaped
          nucleus, often obscured by
          the large quantity of blue-
          black granules in cytoplasm

   Answer: bilobed or S-shaped nucleus, often obscured by the large quantity of blue-black
          granules in cytoplasm
   Type: MA   Page Ref: 383
   Topic: WHITE BLOOD CELLS

17) Column 1: neutrophil
   Column 2: multilobed nucleus; fine
          lilac granules in cytoplasm

   Answer: multilobed nucleus; fine lilac granules in cytoplasm
   Type: MA   Page Ref: 383
   Topic: WHITE BLOOD CELLS

CHAPTER 13   The Cardiovascular System: The Heart

MULTIPLE CHOICE.   Choose the one alternative that best completes the statement or answers the question.

1)   Which of the following is a correct description of the heart?
A) The apex is superior; the base is inferior.
B) The apex is associated with the left side of the heart.
C) The atria are posterior; the ventricles are anterior.
D) A midsagittal section of the body divides the heart into equal right and left portions.

Answer: B
Type: MC   Page Ref: 393
Topic: HEART LOCATION AND SURFACE PROJECTION

2)   The layer of tissue that anchors the heart to the diaphragm and sternum, is fused to the blood vessels entering and leaving the heart, and that prevents overstretching of the heart is the:
A) mediastinum.
B) fibrous pericardium.
C) parietal layer of the serous pericardium.
D) visceral layer of the serous pericardium.

Answer: B
Type: MC   Page Ref: 396
Topic: PERICARDIUM

3)   Desmosomes and communicating junctions are found in:
A) endocardium.   B) chordae tendineae.
C) epicardium.   D) intercalated discs.

Answer: D
Type: MC   Page Ref: 396
Topic: HEART WALL

4)   The _____ is an external feature of the heart that is located between atria and ventricles.
A) fossa ovalis   B) foramen ovale
C) coronary sulcus   D) interventricular sulcus

Answer: C
Type: MC   Page Ref: 397
Topic: CHAMBERS OF THE HEART

5)   The atria of the heart:
A) exhibit ridges in their inner walls called trabeculae carneae.
B) are connected by an opening called the fossa ovalis.
C) contain oxygenated blood.
D) none of the above.

Answer: D
Type: MC   Page Ref: 399
Topic: CHAMBERS OF THE HEART

6)   Which of the following vessels deliver blood to the right atrium?
     1. superior vena cava
     2. inferior vena cava
     3. pulmonary veins
     4. coronary sinus
     5. pulmonary arteries
     A) 1, 2, 3, 4   B) 1, 4, 5   C) 1, 2, 4   D) 2, 3

     Answer: C
     Type: MC   Page Ref: 399
     Topic: CHAMBERS OF THE HEART

7)   Which of the following vessels carries blood away from the heart?
     A) coronary sinus   B) pulmonary vein
     C) inferior vena cava   D) pulmonary artery

     Answer: D
     Type: MC   Page Ref: 399
     Topic: CHAMBERS OF THE HEART

8)   There are four sets of valves in the heart:
     1. bicuspid
     2. tricuspid
     3. pulmonary semilunar
     4. aortic semilunar
     What is the correct order in which blood flows through these valves, starting in the
     right atrium?
     A) 1, 4, 2, 3   B) 1, 3, 2, 4   C) 2, 3, 1, 4   D) 2, 4, 1, 3

     Answer: C
     Type: MC   Page Ref: 403
     Topic: VALVES OF THE HEART

9)   The AV valve in the left side of the heart is known as the:
     A) bicuspid   B) pulmonary semilunar.
     C) tricuspid.   D) aortic semilunar.

     Answer: A
     Type: MC   Page Ref: 403
     Topic: VALVES OF THE HEART

10) Which of the following pairs of terms is most closely matched?
     A) bicuspid valve; oxygenated blood
     B) tricuspid valve; deoxygenated blood
     C) pulmonary semilunar valve; deoxygenated blood
     D) all of the above pairs are correctly matched

     Answer: D
     Type: MC   Page Ref: 402, 403
     Topic: VALVES OF THE HEART

11) Which of the following are branches of the right coronary artery?
   1. marginal branch
   2. circumflex branch
   3. anterior interventricular branch
   4. posterior interventricular branch
   A) 1, 2, 4   B) 1, 3   C) 2, 3   D) 1, 4

   Answer: D
   Type: MC   Page Ref: 406
   Topic: HEART BLOOD SUPPLY

12) Which of the following are branches of the left coronary artery?
   1. marginal branch
   2. circumflex branch
   3. anterior interventricular branch
   4. posterior interventricular branch
   A) 1, 2, 3   B) 2, 4   C) 1, 3   D) 2, 3

   Answer: D
   Type: MC   Page Ref: 406
   Topic: HEART BLOOD SUPPLY

13) Which vessel does *not* supply blood to the right ventricle?
   A) anterior interventricular artery   B) posterior interventricular artery
   C) marginal artery   D) circumflex artery

   Answer: D
   Type: MC   Page Ref: 406
   Topic: HEART BLOOD SUPPLY

14) A vascular sinus, such as the coronary sinus, is best described as:
   A) a large thin-walled vein.
   B) a large thin-walled artery.
   C) a connection between blood vessels that allows for multiple routes of blood flow.
   D) blood-filled connective tissue.

   Answer: A
   Type: MC   Page Ref: 406
   Topic: HEART BLOOD SUPPLY

15) The first part of the heartbeat is due to blood turbulence when the _____ valves close. The second part of the heartbeat is due to blood turbulence when the _____ valves close.
   A) atrioventricular, semilunar   B) semilunar, atrioventricular
   C) tricuspid, bicuspid   D) bicuspid, tricuspid

   Answer: A
   Type: BI   Page Ref: 404, 405
   Topic: HEART SOUNDS

16) The SA node, or pacemaker, of the heart:
   A) is located on the anterior wall of the right atrium.
   B) is located on the anterior wall of the left atrium.
   C) generates action potentials at a faster rate than other regions of the conduction system.
   D) generates action potentials at the same rate as other regions of the conduction system.

   Answer: C
   Type: MC   Page Ref: 407
   Topic: CONDUCTION SYSTEM OF THE HEART

TRUE/FALSE. Write 'T' if the statement is true and 'F' if the statement is false.

1) The pericardial cavity contains pericardial fluid.
   Answer: TRUE
   Type: TF   Page Ref: 396
   Topic: PERICARDIUM

2) The pericardium is composed of elastic connective tissue, to allow for change in volume of the heart.
   Answer: FALSE
   Type: TF   Page Ref: 396
   Topic: PERICARDIUM

3) Desmosomes function to hold cardiac muscle fibers firmly together.
   Answer: TRUE
   Type: TF   Page Ref: 396
   Topic: HEART WALL

4) The fossa ovalis is a thin spot or depression in the interatrial septum.
   Answer: TRUE
   Type: TF   Page Ref: 399
   Topic: CHAMBERS OF THE HEART

5) The coronary sinus carries oxygenated blood.
   Answer: FALSE
   Type: TF   Page Ref: 403
   Topic: CHAMBERS OF THE HEART

6) The pulmonary veins carry deoxygenated blood.
   Answer: FALSE
   Type: TF   Page Ref: 403
   Topic: CHAMBERS OF THE HEART

7) The coronary arteries carry oxygenated blood.
   Answer: TRUE
   Type: TF   Page Ref: 403
   Topic: CHAMBERS OF THE HEART

8) The tricuspid valve is located at the right atrioventricular opening.

Answer: TRUE
Type: TF   Page Ref: 403
Topic: VALVES OF THE HEART

9) Chordae tendineae assist the cusps of the semilunar valves in preventing backflow of blood.

Answer: FALSE
Type: TF   Page Ref: 403
Topic: VALVES OF THE HEART

10) A heart murmur is not always an indication of a serious health problem.

Answer: TRUE
Type: TF   Page Ref: 406
Topic: VALVES OF THE HEART

11) Anastomoses are only found in the heart, and are connections between arteries.

Answer: FALSE
Type: TF   Page Ref: 406
Topic: HEART BLOOD SUPPLY

12) The middle cardiac vein is the main vessel that drains the posterior aspect of the heart.

Answer: TRUE
Type: TF   Page Ref: 406
Topic: HEART BLOOD SUPPLY

13) Cardiac contractions depend upon signals from the autonomic nervous system.

Answer: FALSE
Type: TF   Page Ref: 407
Topic: CONDUCTION SYSTEM OF THE HEART

14) Heart rate is normally altered by the effects of autonomic nervous stimulation or hormones such as thyroid hormone and epinephrine.

Answer: TRUE
Type: TF   Page Ref: 407
Topic: CONDUCTION SYSTEM OF THE HEART

ESSAY. Write your answer in the space provided or on a separate sheet of paper.

1) Surface projection refers to outlining the shape and location of an organ using landmarks on the surface of the organ body. Describe the surface projection of the heart.

Answer: 1. superior right point: 3 cm to the right of the midline, at the superior border of the third right costal cartilage
2. superior left point: 3 cm to the left of the midline, at the inferior border of the second left costal cartilage
(A line connecting these two points corresponds to the base of the heart.)
3. inferior left point: 9 cm to the left of the midline, in the fifth left intercostal space. This corresponds to the apex of the heart.
(A line connecting the superior and inferior left points corresponds to the left border of the heart.)
4. inferior right point: 3 cm to the right of the midline, at the superior border of the sixth right costal cartilage
(A line connecting the inferior left and right points corresponds to the inferior surface of the heart. A line connecting the inferior and superior right points corresponds to the right border of the heart.)
Type: ES   Page Ref: 393, 394
Topic: HEART LOCATION AND SURFACE PROJECTION

2) Trace the path of a drop of blood from the time it enters the heart via the superior vena cava until it is in the arch of the aorta.

Answer: The route described should contain, in order, the following: SVC, right atrium, right ventricle, pulmonary trunk, pulmonary artery, (capillaries in) lungs, pulmonary vein, left atrium, left ventricle, ascending aorta, arch of aorta.
Type: ES   Page Ref: 402, 403
Topic: CHAMBERS OF THE HEART

3) State three functions and four components of the skeleton of the heart.

Answer: The functions are valve and muscle attachment, and electrical insulation between the atria and ventricles. The four components are:
1. the four fibrous rings (right atrioventricular, left atrioventricular, pulmonary, aortic)
2. right fibrous trigone
3. left fibrous trigone
4. conus tendon
Type: ES   Page Ref: 403
Topic: SKELETON OF THE HEART

4) Describe the location and the function of each portion of the conduction system during one cardiac cycle (one complete heart contraction).

Answer: See the list of conduction system components and the description of the spread of excitation on pp. 407.
Type: ES   Page Ref: 407
Topic: CONDUCTION SYSTEM OF THE HEART

SHORT ANSWER. Write the word or phrase that best completes each statement or answers the question.

1) The study of the heart and heart disease is called _____.

Answer: cardiology
Type: SA   Page Ref: 393
Topic: INTRODUCTION

2) The pericardial cavity is found between the epicardium and the _____ layer of the serous pericardium.

Answer: parietal
Type: SA   Page Ref: 396
Topic: PERICARDIUM

3) Each atrium has a/an _____ that functions to increase the effective volume of the atrium.

Answer: auricle
Type: SA   Page Ref: 397
Topic: CHAMBERS OF THE HEART

4) The anterior and posterior interventricular sulci are external features that mark the location of the _____ inside the heart.

Answer: interventricular septum
Type: SA   Page Ref: 397, 399
Topic: CHAMBERS OF THE HEART

5) AV valve cusps are connected by strong cords called _____ to _____ muscles on the walls of the ventricles.

Answer: chordae tendineae, papillary
Type: SA   Page Ref: 403
Topic: VALVES OF THE HEART

6) The _____ branch is the artery located in the coronary sulcus.

Answer: circumflex
Type: SA   Page Ref: 406
Topic: HEART BLOOD SUPPLY

7) The main vein draining the anterior aspect of the heart and emptying into the coronary sinus is the _____ vein.

Answer: great cardiac
Type: SA   Page Ref: 406
Topic: HEART BLOOD SUPPLY

8) The cells that make up the conduction system of the heart are specialized _____ cells.

Answer: cardiac muscle
Type: SA   Page Ref: 407
Topic: CONDUCTION SYSTEM OF THE HEART

MATCHING. Choose the item in Column 2 that best matches each item in Column 1.

Match the layers of the heart wall in Column 1 with their description in Column 2.

1) Column 1: epicardium
    Column 2: composed of mesothelium
              and connective tissue

    Answer: composed of mesothelium and connective tissue
    Type: MA   Page Ref: 396
    Topic: HEART WALL

2) Column 1: myocardium
    Column 2: thickest layer, muscular
    Foil: found only in walls of
               ventricles

    Answer: thickest layer, muscular
    Type: MA   Page Ref: 396
    Topic: HEART WALL

3) Column 1: endocardium
    Column 2: continuous with the lining
              of the blood vessels

    Answer: continuous with the lining of the blood vessels
    Type: MA   Page Ref: 396
    Topic: HEART WALL

4) Column 1: endocardium
    Column 2: covers the valves

    Answer: covers the valves
    Type: MA   Page Ref: 396
    Topic: HEART WALL

5) Column 1: epicardium
    Column 2: layer next to pericardial
              cavity

    Answer: layer next to pericardial cavity
    Type: MA   Page Ref: 396
    Topic: HEART WALL

Match the components of the conduction system of the heart in Column 1 with the descriptions in Column 2.

6) Column 1: sinoatrial node
    Column 2: pacemaker

    Answer: pacemaker
    Type: MA   Page Ref: 407
    Topic: CONDUCTION SYSTEM OF THE HEART

7)   Column 1: sinoatrial node
        Column 2: sends wave of
                depolarization throughout
                atria

        Answer: sends wave of depolarization throughout atria
        Type: MA   Page Ref: 407
        Topic: CONDUCTION SYSTEM OF THE HEART

8)   Column 1: atrioventricular node
        Column 2: located in the interatrial
                septum

        Answer: located in the interatrial septum
        Type: MA   Page Ref: 407
        Topic: CONDUCTION SYSTEM OF THE HEART

9)   Column 1: atrioventricular bundle
        Column 2: the electrical connection
                between the atria and the
                ventricles

        Answer: the electrical connection between the atria and the ventricles
        Type: MA   Page Ref: 408
        Topic: CONDUCTION SYSTEM OF THE HEART

10)   Column 1: bundle branches
        Column 2: carry depolarization wave
                toward apex of heart

        Answer: carry depolarization wave toward apex of heart
        Type: MA   Page Ref: 408
        Topic: CONDUCTION SYSTEM OF THE HEART

11)   Column 1: conduction myofibers
        Column 2: cause actual stimulation of
                cardiac muscle fibers in the
                ventricles

        Answer: cause actual stimulation of cardiac muscle fibers in the ventricles
        Type: MA   Page Ref: 408
        Topic: CONDUCTION SYSTEM OF THE HEART

Match the vessels in Column 1 with their descriptions in Column 2.
12)   Column 1: pulmonary trunk
        Column 2: carries deoxygenated blood
                away from the heart
        Foil: carries deoxygenated blood
                toward the heart

        Answer: carries deoxygenated blood away from the heart
        Type: MA   Page Ref: 403
        Topic: OVERVIEW OF CIRCULATION

13)   Column 1: aorta
       Column 2: carries oxygenated blood
                 away from the heart

       Answer: carries oxygenated blood away from the heart
       Type: MA   Page Ref: 403
       Topic: OVERVIEW OF CIRCULATION

14)   Column 1: pulmonary vein
       Column 2: carries oxygenated blood
                 toward the heart

       Answer: carries oxygenated blood toward the heart
       Type: MA   Page Ref: 403
       Topic: OVERVIEW OF CIRCULATION

15)   Column 1: pulmonary vein
       Column 2: enters left atrium
       Foil: enters right atrium

       Answer: enters left atrium
       Type: MA   Page Ref: 403
       Topic: CHAMBERS OF THE HEART

16)   Column 1: pulmonary trunk
       Column 2: connected to right ventricle

       Answer: connected to right ventricle
       Type: MA   Page Ref: 403
       Topic: CHAMBERS OF THE HEART

17)   Column 1: aorta
       Column 2: connected to left ventricle

       Answer: connected to left ventricle
       Type: MA   Page Ref: 403
       Topic: CHAMBERS OF THE HEART

Match the ventricles of the heart in Column 1 with their descriptions in Column 2.

18)   Column 1: right ventricle
       Column 2: receives deoxygenated blood

       Answer: receives deoxygenated blood
       Type: MA   Page Ref: 399
       Topic: OVERVIEW OF CIRCULATION

19)   Column 1: left ventricle
       Column 2: pumps blood into the aorta

       Answer: pumps blood into the aorta
       Type: MA   Page Ref: 400
       Topic: OVERVIEW OF CIRCULATION

20) Column 1: left ventricle
Column 2: thicker myocardium
Foil: pumps a larger volume of blood

Answer: thicker myocardium
Type: MA   Page Ref: 401
Topic: CHAMBERS OF THE HEART

21) Column 1: right ventricle
Column 2: pumps blood to the lungs

Answer: pumps blood to the lungs
Type: MA   Page Ref: 399
Topic: CHAMBERS OF THE HEART

22) Column 1: left ventricle
Column 2: entrance into this ventricle is guarded by the bicuspid valve

Answer: entrance into this ventricle is guarded by the bicuspid valve
Type: MA   Page Ref: 400
Topic: CHAMBERS OF THE HEART

23) Column 1: right ventricle
Column 2: pumps blood through the pulmonary semilunar valve
Foil: pumps blood through the tricuspid valve

Answer: pumps blood through the pulmonary semilunar valve
Type: MA   Page Ref: 399
Topic: CHAMBERS OF THE HEART

CHAPTER 14   The Cardiovascular System: Blood Vessels

MULTIPLE CHOICE. Choose the one alternative that best completes the statement or answers the question.

1)   Epithelial tissue forms part of which layer(s) of an artery wall?
     1. tunica interna
     2. tunica media
     3. tunica externa
     A) 1, 2, 3   B) 1 only   C) 1, 3   D) 1, 2

     Answer: B
     Type: MC   Page Ref: 417
     Topic: ANATOMY OF BLOOD VESSELS

2)   Elastic arteries:
     A) are medium-sized arteries.
     B) restrict blood flow due to the force needed to stretch their walls.
     C) are also called distributing arteries.
     D) include the largest diameter arteries in the body.

     Answer: D
     Type: MC   Page Ref: 417
     Topic: ANATOMY OF BLOOD VESSELS

3)   Which of the following is a muscular artery?
     A) axillary   B) common iliac
     C) common carotid   D) aorta

     Answer: B
     Type: MC   Page Ref: 417
     Topic: ANATOMY OF BLOOD VESSELS

4)   The layer(s) of an artery wall that is/are responsible for vasoconstriction and vasodilation is/are the:
     A) tunica externa.   B) tunica interna.
     C) tunica media.   D) all of the above.

     Answer: C
     Type: MC   Page Ref: 417
     Topic: ANATOMY OF BLOOD VESSELS

5)   The thinnest-walled blood vessels in the body are:
     A) capillaries.   B) veins.   C) venules.   D) arterioles.

     Answer: A
     Type: MC   Page Ref: 419
     Topic: ANATOMY OF BLOOD VESSELS

6)   Abundant capillary networks are located in the:
     A) epidermis of the skin.   B) lens of the eye.
     C) lining of the stomach.   D) liver.

     Answer: D
     Type: MC   Page Ref: 419
     Topic: ANATOMY OF BLOOD VESSELS

7) Blood flows through the following vessels in what order?
   1. arteriole
   2. capillary
   3. thoroughfare channel
   4. metarteriole
   5. venule
   A) 1, 2, 4, 5   B) 1, 4, 2, 5   C) 5, 2, 3, 1   D) 1, 2, 3, 4

   Answer: B
   Type: MC   Page Ref: 419
   Topic: ANATOMY OF BLOOD VESSELS

8) Materials move between capillary blood and interstitial fluid through:
   A) fenestrations.   B) intercellular clefts.
   C) pinocytic vesicles.   D) all of the above.

   Answer: D
   Type: MC   Page Ref: 420
   Topic: ANATOMY OF BLOOD VESSELS

9) Stellate reticuloendothelial cells are specialized lining cells that are found in _____ in the _____.
   A) arterioles, brain   B) sinusoids, liver
   C) choroid plexuses, brain   D) capillaries, endocrine glands

   Answer: B
   Type: MC   Page Ref: 421
   Topic: ANATOMY OF BLOOD VESSELS

10) A difference between veins and arteries is:
    A) arteries have three coats; veins have only two coats.
    B) venous blood is under more pressure than is arterial blood.
    C) arteries have their own blood supply (vasa vasorum); veins do not.
    D) the tunica interna and tunica media are thinner in veins than in arteries.

    Answer: D
    Type: MC   Page Ref: 421
    Topic: ANATOMY OF BLOOD VESSELS

11) The _____ contain(s) the largest volume of blood when the body is at rest.
    A) arteries and arterioles   B) veins and venules
    C) heart   D) systemic capillaries

    Answer: B
    Type: MC   Page Ref: 422
    Topic: ANATOMY OF BLOOD VESSELS

12) Blood in systemic circulation travels from the _____, then through arteries, capillaries, and veins, to the _____.
    A) left ventricle, right atrium   B) left atrium, right ventricle
    C) left atrium, right atrium   D) right ventricle, left atrium

    Answer: A
    Type: BI   Page Ref: 417–421
    Topic: ANATOMY OF BLOOD VESSELS

13) Blood in pulmonary circulation travels from the _____, then through arteries, capillaries, and veins, to the _____.
A) left ventricle, right atrium
B) left atrium, right ventricle
C) left atrium, right atrium
D) right ventricle, left atrium

Answer: D
Type: MC   Page Ref: 466
Topic: PULMONARY CIRCULATION

14) The branches of the arch of the aorta, in correct order, are:
1. brachiocephalic trunk
2. left common carotid
3. left subclavian
A) 1, 2, 3   B) 1, 3, 2   C) 2, 1, 3   D) 3, 2, 1

Answer: A
Type: MC   Page Ref: 425
Topic: SYSTEMIC CIRCULATION

15) What is the correct route that a drop of blood would follow as it flows through the following vessels?
1. inferior vena cava
2. thoracic aorta
3. renal artery
4. abdominal aorta
5. renal vein
A) 1, 3, 5, 4, 2   B) 2, 3, 5, 4, 1   C) 2, 3, 4, 5, 1   D) 2, 4, 3, 5, 1

Answer: D
Type: MC   Page Ref: 423, 425, 459
Topic: SYSTEMIC CIRCULATION

16) What is the correct route that a drop of blood would follow as it flows through the following vessels?
1. external iliac artery
2. popliteal artery
3. anterior tibial artery
4. common iliac artery
5. femoral artery
6. dorsalis pedis artery
A) 1, 4, 5, 3, 6, 2   B) 4, 1, 5, 2, 3, 6   C) 4, 1, 2, 5, 3, 6   D) 6, 3, 2, 1, 5, 4

Answer: B
Type: MC   Page Ref: 443, 444
Topic: SYSTEMIC CIRCULATION

17) What is the correct route that a drop of blood would follow as it flows through the following vessels?
1. superior vena cava
2. internal jugular vein
3. sigmoid sinuses
4. subclavian vein
5. brachiocephalic vein

A) 3, 2, 4, 5, 1   B) 1, 2, 5, 3, 4   C) 2, 3, 1, 4, 5   D) 3, 2, 5, 4, 1

Answer: A
Type: MC   Page Ref: 450
Topic: SYSTEMIC CIRCULATION

18) What is the correct route that a drop of blood would follow as it flows through the following vessels?
1. ulnar vein
2. palmar venous arch
3. brachial vein
4. subclavian vein
5. axillary vein
6. brachiocephalic vein

A) 2, 1, 3, 5, 4, 6   B) 2, 1, 3, 4, 5, 6   C) 2, 3, 1, 5, 4, 6   D) 6, 5, 4, 3, 1, 2

Answer: A
Type: MC   Page Ref: 453
Topic: SYSTEMIC CIRCULATION

19) Which of the following do/does *not* provide blood for the cerebral arterial circle?
A) internal carotid arteries   B) vertebral arteries
C) external carotid arteries   D) basilar artery

Answer: C
Type: MC   Page Ref: 443
Topic: SYSTEMIC CIRCULATION

20) The gastric, pancreatic, and splenic veins are part of the:
A) systemic circulation   B) hepatic portal circulation
C) pulmonary circulation   D) fetal circulation only

Answer: B
Type: MC   Page Ref: 466
Topic: HEPATIC PORTAL CIRCULATION

21) Which of the following is *not* a part of the hepatic portal system?
A) splenic vein   B) superior mesenteric vein
C) hepatic veins   D) inferior mesenteric vein

Answer: C
Type: MC   Page Ref: 466
Topic: HEPATIC PORTAL CIRCULATION

22) Pulmonary circulation differs from systemic circulation in that:
A) blood passes first through veins, then capillaries, then arteries.
B) arteries carry deoxygenated blood and veins carry oxygenated blood.
C) all of the blood in pulmonary circulation is oxygenated blood.
D) low oxygen levels cause pulmonary vessels to dilate, but cause systemic vessels to constrict.

Answer: B
Type: MC   Page Ref: 466
Topic: PULMONARY CIRCULATION

23) Trace the path of a drop of blood through pulmonary circulation, starting as it leaves the heart.
1. left atrium
2. right ventricle
3. pulmonary vein
4. pulmonary artery
5. pulmonary capillaries
6. pulmonary trunk
A) 1, 6, 3, 5, 4, 2   B) 1, 6, 4, 5, 3, 2   C) 2, 6, 3, 5, 4, 1   D) 2, 6, 4, 5, 3, 1

Answer: D
Type: MC   Page Ref: 466
Topic: PULMONARY CIRCULATION

24) Which of the following is *not* considered a change in the cardiovascular system that is due to aging?
A) increased cholesterol levels and atherosclerosis
B) decreased elasticity of the aorta
C) decreased number of cardiac muscle cells
D) a decline in maximum heart rate

Answer: C
Type: MC   Page Ref: 472
Topic: CARDIOVASCULAR SYSTEM: Aging

TRUE/FALSE.   Write 'T' if the statement is true and 'F' if the statement is false.

1) A tunica interna of endothelium is found only in arteries.
Answer: FALSE
Type: TF   Page Ref: 417–419
Topic: ANATOMY OF BLOOD VESSELS

2) In an artery wall, the tunica media contains smooth muscle and elastic fibers.
Answer: TRUE
Type: TF   Page Ref: 417
Topic: ANATOMY OF BLOOD VESSELS

3) A blockage in an end artery is more serious than in an anastomosing artery.
Answer: TRUE
Type: TF   Page Ref: 419
Topic: ANATOMY OF BLOOD VESSELS

4) At the level of the fourth lumbar vertebra, the aorta divides into left and right common iliac arteries.

Answer: TRUE
Type: TF   Page Ref: 437
Topic: SYSTEMIC CIRCULATION

5) The blood that sustains lung tissue is supplied by the systemic circulation.

Answer: TRUE
Type: TF   Page Ref: 435
Topic: SYSTEMIC CIRCULATION

6) The SVC and IVC are both located to the right of the midline of the body.

Answer: TRUE
Type: TF   Page Ref: 448, 449
Topic: SYSTEMIC CIRCULATION

7) Pulmonary circulation carries blood between capillaries in the heart and capillaries in the lungs.

Answer: FALSE
Type: TF   Page Ref: 466
Topic: PULMONARY CIRCULATION

8) Blood in an umbilical artery travels from the fetus to the placenta.

Answer: TRUE
Type: TF   Page Ref: 468
Topic: FETAL CIRCULATION

ESSAY.   Write your answer in the space provided or on a separate sheet of paper.

1) Describe the three layers that make up the wall of an artery. Using this as a standard, compare artery walls to capillary and vein walls.

Answer: Summary: The three layers of an artery wall are described on pp. 417. Capillaries consist of endothelium and basement membrane only. Large veins contain all three layers, but are thinner walled than arteries, with less elastic and muscle tissue. Small veins lack the tunica media.
Type: ES   Page Ref: 420
Topic: ANATOMY OF BLOOD VESSELS

2) Is the aorta an elastic or a muscular artery? Compare the structure of the wall of the aorta to an artery that is of the other type.

Answer: The aorta is an elastic (conducting) artery. It contains a large quantity of elastic fiber in the tunica media. The wall stretches and recoils to assist the movement of blood. Muscular (distributing) arteries, on the other hand, have larger amounts of smooth muscle in the tunica media, and are relatively thicker walled vessels. They are capable of vasodilation and vasoconstriction.
Type: ES   Page Ref: 417
Topic: ANATOMY OF BLOOD VESSELS

3) Diagram a portion of a capillary network, labeling the features that are found between the arteriole and the venule. Do not draw the smooth muscle fibers, but indicate on your drawing the regions where they are found. Label the zone of oxygenated blood. State the function of each portion or feature of the capillary network that you have labeled.

Answer: The diagram should be similar to Fig. 14.3. The arteriole supplies oxygenated blood, which is delivered to the capillary by a branch called a metarteriole. When the smooth muscle precapillary sphincters are relaxed, blood will flow into a true capillary from the metarteriole, or sometimes directly from an arteriole. Venules receive the deoxygenated blood from the capillaries. There exists a thoroughfare route from arteriole to venule, through which the blood will flow when the precapillary sphincters are closed. Smooth muscle is present in the walls of the arterioles, and proximal metarterioles to control blood flow.

Type: ES   Page Ref: 420
Topic: ANATOMY OF BLOOD VESSELS

4) Define portal system. Explain how the hepatic portal system fits your definition and describe its function.

Answer: A portal system carries blood from one capillary network to another; in this case, from capillaries of the spleen, stomach, pancreas, intestines, and gall bladder to sinusoids of the liver via the hepatic portal vein. This allows the liver to process the nutrient-rich blood from the gastrointestinal tract and to remove bacteria and toxic substances.

Type: ES   Page Ref: 466
Topic: HEPATIC PORTAL CIRCULATION

SHORT ANSWER. Write the word or phrase that best completes each statement or answers the question.

1) A decrease in the diameter of a blood vessel due to the contraction of smooth muscle in the vessel wall is called _____.

Answer: vasoconstriction
Type: SA   Page Ref: 417
Topic: ANATOMY OF BLOOD VESSELS

2) The thickest layer of an artery wall is the tunica _____.

Answer: media
Type: SA   Page Ref: 417
Topic: ANATOMY OF BLOOD VESSELS

3) The hollow center of a blood vessel is called the _____.

Answer: lumen
Type: SA   Page Ref: 417
Topic: ANATOMY OF BLOOD VESSELS

4) Collateral circulation occurs for any tissue that receives blood from two or more arteries that are joined by connecting blood vessels called _____.

Answer: anastomoses
Type: SA   Page Ref: 419
Topic: ANATOMY OF BLOOD VESSELS

5) Nonanastomosing arteries are called _____ arteries.

Answer: end
Type: SA   Page Ref: 419
Topic: ANATOMY OF BLOOD VESSELS

6) The ring of smooth muscle that controls the flow of blood entering a capillary is the _____.

Answer: precapillary sphincter
Type: SA   Page Ref: 419
Topic: ANATOMY OF BLOOD VESSELS

7) Before reaching a capillary, blood from an artery must flow through a vessel called a/an _____.

Answer: arteriole
Type: SA   Page Ref: 419
Topic: ANATOMY OF BLOOD VESSELS

8) Extensions of the tunica interna of veins form _____ that help prevent the backflow of blood.

Answer: valves
Type: SA   Page Ref: 421
Topic: ANATOMY OF BLOOD VESSELS

9) Vascular sinuses contain _____ (oxygenated or deoxygenated) blood.

Answer: deoxygenated
Type: SA   Page Ref: 421
Topic: ANATOMY OF BLOOD VESSELS

10) Two main blood reservoirs of the body are the veins in the _____ and _____.

Answer: abdominal organs, skin
Type: SA   Page Ref: 422
Topic: ANATOMY OF BLOOD VESSELS

11) The first two vessels to branch from the aorta are the _____.

Answer: right and left coronary arteries
Type: SA   Page Ref: 425
Topic: SYSTEMIC CIRCULATION

12) The superior mesenteric artery branches from the _____.

Answer: abdominal aorta
Type: SA   Page Ref: 437
Topic: SYSTEMIC CIRCULATION

13) The largest diameter vein in the body is the _____.

Answer: inferior vena cava
Type: SA   Page Ref: 448
Topic: SYSTEMIC CIRCULATION

14) The external iliac vein is formed by the union of the _____ and the _____.

Answer: femoral, great saphenous
Type: SA   Page Ref: 463
Topic: SYSTEMIC CIRCULATION

15) The umbilical cord contains three blood vessels: one umbilical _____ and two umbilical _____.

Answer: vein, arteries
Type: SA   Page Ref: 468
Topic: FETAL CIRCULATION

16) In the fetus, the ductus _____ allows most blood to bypass the lungs.

Answer: arteriosus
Type: SA   Page Ref: 468
Topic: FETAL CIRCULATION

17) The superior mesenteric vein empties into the _____.

Answer: hepatic portal vein
Type: SA   Page Ref: 466
Topic: HEPATIC PORTAL CIRCULATION

MATCHING. Choose the item in Column 2 that best matches each item in Column 1.

Match the divisions of the aorta in Column 1 with vessels that branch directly from them in Column 2.

1) Column 1: arch of aorta
Column 2: left subclavian artery
Foil: right subclavian artery

Answer: left subclavian artery
Type: MA   Page Ref: 425
Topic: SYSTEMIC CIRCULATION

2) Column 1: thoracic aorta
Column 2: intercostal arteries

Answer: intercostal arteries
Type: MA   Page Ref: 425
Topic: SYSTEMIC CIRCULATION

3) Column 1: abdominal aorta
Column 2: renal arteries

Answer: renal arteries
Type: MA   Page Ref: 425
Topic: SYSTEMIC CIRCULATION

4) Column 1: ascending aorta
Column 2: right coronary artery
Foil: pulmonary artery

Answer: right coronary artery
Type: MA   Page Ref: 425
Topic: SYSTEMIC CIRCULATION

Match the arteries in Column 1 with their descriptions in Column 2.

5)   Column 1: radial artery
      Column 2: its branches contribute to
              the palmar arches
      Foil: located in medial aspect of
              forearm

      Answer: its branches contribute to the palmar arches
      Type: MA   Page Ref: 429
      Topic: SYSTEMIC CIRCULATION

6)   Column 1: vertebral artery
      Column 2: located in cervical
              transverse foramina

      Answer: located in cervical transverse foramina
      Type: MA   Page Ref: 430
      Topic: SYSTEMIC CIRCULATION

7)   Column 1: axillary artery
      Column 2: blood flows from this vessel
              directly into the brachial
              artery

      Answer: blood flows from this vessel directly into the brachial artery
      Type: MA   Page Ref: 429
      Topic: SYSTEMIC CIRCULATION

8)   Column 1: brachiocephalic trunk
      Column 2: gives rise to right
              subclavian artery

      Answer: gives rise to right subclavian artery
      Type: MA   Page Ref: 429
      Topic: SYSTEMIC CIRCULATION

9)   Column 1: cerebral arterial circle
      Column 2: receives blood that has
              been delivered to the head
              by internal carotid and
              vertebral arteries

      Answer: receives blood that has been delivered to the head by internal carotid and
              vertebral arteries
      Type: MA   Page Ref: 430
      Topic: SYSTEMIC CIRCULATION

10)   Column 1: basilar artery
        Column 2: formed by union of right
                    and left vertebral arteries
        Foil: located on the medial
                    aspect of the forearm and
                    arm

Answer: formed by union of right and left vertebral arteries
Type: MA   Page Ref: 430
Topic: SYSTEMIC CIRCULATION

Match the abdominal arteries in Column 1 with the structures or organs they supply in Column 2.

11)   Column 1: renal artery
        Column 2: kidney
        Foil: gonads

Answer: kidney
Type: MA   Page Ref: 438
Topic: SYSTEMIC CIRCULATION

12)   Column 1: inferior mesenteric artery
        Column 2: large intestine
        Foil: small intestine

Answer: large intestine
Type: MA   Page Ref: 438
Topic: SYSTEMIC CIRCULATION

13)   Column 1: splenic artery
        Column 2: pancreas, spleen, and
                    stomach

Answer: pancreas, spleen, and stomach
Type: MA   Page Ref: 437
Topic: SYSTEMIC CIRCULATION

14)   Column 1: lumbar arteries
        Column 2: spinal cord and meninges
        Foil: sacrum

Answer: spinal cord and meninges
Type: MA   Page Ref: 438
Topic: SYSTEMIC CIRCULATION

15)   Column 1: inferior phrenic artery
        Column 2: diaphragm
        Foil: testes and ovaries

Answer: diaphragm
Type: MA   Page Ref: 438
Topic: SYSTEMIC CIRCULATION

16)   Column 1: common hepatic artery
        Column 2: liver, stomach, and
                  duodenum

        Answer: liver, stomach, and duodenum
        Type: MA   Page Ref: 437
        Topic: SYSTEMIC CIRCULATION

Match the veins in Column 1 with the structures or organs they drain in Column 2.
17)   Column 1: external jugular vein
        Column 2: parotid gland, scalp
        Foil: brain

        Answer: parotid gland, scalp
        Type: MA   Page Ref: 450
        Topic: SYSTEMIC CIRCULATION

18)   Column 1: basilic vein
        Column 2: medial aspect of forearm
                  and arm
        Foil: lateral aspect of forearm
                   and arm

        Answer: medial aspect of forearm and arm
        Type: MA   Page Ref: 453
        Topic: SYSTEMIC CIRCULATION

19)   Column 1: internal iliac vein
        Column 2: buttocks, urinary bladder,
                  uterus, prostate gland

        Answer: buttocks, urinary bladder, uterus, prostate gland
        Type: MA   Page Ref: 459
        Topic: SYSTEMIC CIRCULATION

20)   Column 1: suprarenal vein
        Column 2: adrenal gland

        Answer: adrenal gland
        Type: MA   Page Ref: 459
        Topic: SYSTEMIC CIRCULATION

21)   Column 1: great saphenous vein
        Column 2: medial aspect of leg, thigh,
                  and foot

        Answer: medial aspect of leg, thigh, and foot
        Type: MA   Page Ref: 462
        Topic: SYSTEMIC CIRCULATION

22)   Column 1: hepatic veins
        Column 2: liver
        Foil: diaphragm

        Answer: liver
        Type: MA   Page Ref: 459
        Topic: SYSTEMIC CIRCULATION

23) Column 1: anterior tibial
Column 2: foot

Answer: foot
Type: MA   Page Ref: 463
Topic: SYSTEMIC CIRCULATION

CHAPTER 15   The Lymphatic System

MULTIPLE CHOICE. Choose the one alternative that best completes the statement or answers the question.

1)   Lymphatic vessels in the skin follow veins. Lymphatic vessels in the viscera follow arteries.
      A) Both statements are true.
      B) Both statements are false.
      C) The first statement is true; the second is false.
      D) The second statement is true; the first is false.

      Answer: A
      Type: MC   Page Ref: 478
      Topic: LYMPHATIC VESSELS

2)   Lymphatic capillaries are found in tissues throughout the body except for:
      A) red bone marrow.   B) brain.
      C) spinal cord.   D) all of the above.

      Answer: D
      Type: MC   Page Ref: 478
      Topic: LYMPHATIC VESSELS

3)   Which of the following occurs as a result of the presence of lacteals in the wall of the intestine?
      A) Most dietary fats do not enter the hepatic portal circulation.
      B) The lymph receives digested milk sugars, which gives the lymph its creamy color.
      C) All end products of digestion are carried by lymph.
      D) Lymph enters the hepatic portal circulation.

      Answer: A
      Type: MC   Page Ref: 478
      Topic: LYMPHATIC VESSELS

4)   The composition of lymph most closely resembles:
      A) intracellular fluid.   B) plasma.
      C) interstitial fluid.   D) synovial fluid.

      Answer: C
      Type: MC   Page Ref: 480
      Topic: LYMPH

5) A correct sequence for fluid flow in the body is:
    A) arteries, lymphatic capillaries, lymphatic ducts, heart, veins, blood capillaries, interstitial spaces.
    B) blood capillaries, interstitial spaces, lymphatic capillaries, lymphatic ducts, veins, heart, arteries.
    C) lymphatic capillaries, interstitial spaces, lymphatic ducts, veins, heart, arteries, blood capillaries.
    D) veins, heart, arteries, lymphatic ducts, interstitial spaces, lymphatic capillaries, blood capillaries.
    Answer: B
    Type: MC   Page Ref: 478
    Topic: LYMPH

6) Which of the following is *not* an important factor that assists the flow of lymph through lymphatic vessels?
    A) valves in lymphatic vessels   B) skeletal muscle contractions
    C) gravity   D) respiratory movements
    Answer: C
    Type: MC   Page Ref: 480
    Topic: LYMPH

7) The right lymphatic duct receives lymph from all of the following except the:
    A) right lung.   B) right side of heart.
    C) right kidney.   D) right side of head and neck.
    Answer: C
    Type: MC   Page Ref: 481
    Topic: LYMPHATIC VESSELS

8) The thoracic duct receives lymph from all of the following *except*:
    A) both lower limbs.   B) the right upper limb.
    C) the stomach.   D) the left side of head and neck.
    Answer: B
    Type: MC   Page Ref: 481
    Topic: LYMPHATIC VESSELS

9) The lining of the respiratory, digestive, reproductive, and urinary tracts contains lymphatic tissue of the type called:
    A) lymphatic nodules.   B) lymph nodes.
    C) lacteals.   D) white pulp.
    Answer: A
    Type: MC   Page Ref: 483
    Topic: LYMPHATIC TISSUES

10) Bone marrow and the thymus gland are called the primary lymphatic organs because they:
A) filter all lymph.
B) produce B and T cells.
C) are functional only in primary development, i.e., in the fetus only.
D) None of the above. They are *not* primary lymphatic organs. The spleen is the only primary lymphatic organ.

Answer: B
Type: MC   Page Ref: 482
Topic: LYMPHATIC TISSUES

11) Lymphatic organs that have a hilus are the:
A) spleen and lymph nodes.   B) tonsils and appendix.
C) tonsils and lymph nodes.   D) spleen and thymus gland.

Answer: A
Type: MC   Page Ref: 484
Topic: LYMPHATIC TISSUES

12) Which of the following statements about lymph nodes is *not* true?
A) Afferent lymphatic vessels direct lymph out of the lymph node, along the convex surface.
B) The efferent lymphatic vessel is attached at the hilus, along with blood vessels.
C) The cortex consists of follicles that resemble lymphatic nodules.
D) Cells found in the lymph nodes include T cells, B cells, and macrophages.

Answer: A
Type: MC   Page Ref: 484
Topic: LYMPHATIC TISSUES

13) The parenchyma of the spleen consists of:
A) cortex and medulla.   B) B cells and T cells.
C) lobes and lobules.   D) red pulp and white pulp.

Answer: D
Type: MC   Page Ref: 484
Topic: LYMPHATIC TISSUES

14) Which of the following vessels does *not* exist for the spleen?
A) splenic artery   B) splenic vein
C) afferent lymphatic vessel   D) efferent lymphatic vessel

Answer: C
Type: MC   Page Ref: 484
Topic: LYMPHATIC TISSUES

15) The lymphatic vessels arise from:
A) ectoderm.   B) endoderm.
C) mesoderm.   D) all of the above.

Answer: C
Type: MC   Page Ref: 496
Topic: DEVELOPMENTAL ANATOMY OF THE LYMPHATIC SYSTEM

TRUE/FALSE. Write 'T' if the statement is true and 'F' if the statement is false.

1) Lymphatic vessels are not located in tissues that lack blood capillaries.

Answer: TRUE
Type: TF   Page Ref: 478
Topic: LYMPHATIC VESSELS

2) Lymphatic capillaries merge to form larger veinlike vessels called lymphatic vessels.

Answer: TRUE
Type: TF   Page Ref: 478
Topic: LYMPHATIC VESSELS

3) Edema, or excess accumulation of interstitial fluid, is due to anchoring filaments causing closure of the spaces between lymphatic endothelial cells.

Answer: FALSE
Type: TF   Page Ref: 480
Topic: LYMPHATIC VESSELS

4) Within the lymphatic system, lymph flows, in order, through the following structures: lymphatic capillaries, lymphatic vessels, lymph nodes, lymph trunks, lymphatic ducts.

Answer: TRUE
Type: TF   Page Ref: 480, 481
Topic: LYMPHATIC VESSELS

5) Tonsils are examples of lymphatic nodules.

Answer: TRUE
Type: TF   Page Ref: 486
Topic: LYMPHATIC TISSUES

6) The three groups of nodes called parotid, buccal, and submandibular make up the facial group of nodes.

Answer: FALSE
Type: TF   Page Ref: 487
Topic: LYMPHATIC TISSUES

7) Each lobe of the thymus gland consists of an outer cortex and an inner medulla.

Answer: FALSE
Type: TF   Page Ref: 483
Topic: LYMPHATIC TISSUES

8) The thymus gland reaches its maximum size around 10 to 12 years of age, after which it begins to atrophy.

Answer: TRUE
Type: TF   Page Ref: 483
Topic: LYMPHATIC TISSUES

9) The stroma of the spleen consists of the capsule, trabeculae, reticular fibers, the red pulp, and the white pulp.

Answer: FALSE
Type: TF   Page Ref: 484
Topic: LYMPHATIC TISSUES

10) The lymphatic vessels arise from the same type of embryonic tissue as do veins.

Answer: TRUE
Type: TF   Page Ref: 496
Topic: DEVELOPMENTAL ANATOMY OF THE LYMPHATIC SYSTEM

ESSAY. Write your answer in the space provided or on a separate sheet of paper.

1) List three functions of the lymphatic system and name the components of the lymphatic system primarily responsible for each.

Answer: 1. draining of ISF by lymphatic vessels
2. transportation of lipids and lipid-soluble vitamins by lymphatic vessels from the GI tract to the blood
3. protection against specific invaders via lymphocytes
Type: ES   Page Ref: 478
Topic: INTRODUCTION

2) Name and describe the location of the two main lymphatic ducts. Describe the portions of the body drained by each.

Answer: The thoracic duct, located left of the midline in the thorax, receives lymph from the cysterna chyli anterior to the second lumbar vertebra. It receives lymph from the entire body below the ribs and the left side of the head, neck, chest, and left upper limb. It drains into the left subclavian vein.
The right lymphatic duct, located right of the midline in the thorax, drains the upper right region of the body via the right jugular, right subclavian, and right bronchomediastinal trunk.
Type: ES   Page Ref: 415
Topic: LYMPHATIC VESSELS

3) List the nodes that make up the axillary group and state the region of the body drained by each.

Answer: 1. Lateral nodes drain most of the upper limb.
2. Pectoral nodes drain the anterior and lateral thoracic wall, and the central and lateral portion of the mammary glands.
3. Subscapular nodes drain the posterior part of the neck and thoracic wall.
4. Central nodes drain the lateral, pectoral, and subscapular nodes.
5. Subclavicular nodes drain the deltopectoral nodes, which drain the upper limb.
Type: ES   Page Ref: 489
Topic: LYMPHATIC TISSUES

4) Lymph nodes of the body are arranged into five principal groups. Name any one group of lymph nodes and describe its location and the region of the body it drains.

Answer: Answers will vary depending upon choice of nodes.
        Details in Exhibits 15.1–15.5.
Type: ES   Page Ref: 487–494
Topic: LYMPHATIC TISSUES

SHORT ANSWER. Write the word or phrase that best completes each statement or answers the question.

1) Lymphatic capillaries differ from blood capillaries in that they are closed-ended and they have a slightly _____ (larger or smaller) diameter than blood capillaries.

Answer: larger
Type: SA   Page Ref: 478
Topic: LYMPHATIC VESSELS

2) Fluid enters lymphatic capillaries when the pressure in the interstitial fluid spaces is _____ (greater or less) than the pressure inside the lymphatic capillary.

Answer: greater
Type: SA   Page Ref: 480
Topic: LYMPHATIC VESSELS

3) Lymphatic capillaries in intestinal villi are called _____.

Answer: lacteals
Type: SA   Page Ref: 478
Topic: LYMPHATIC VESSELS

4) Breathing assists the flow of lymph; lymph flows from the abdominal region to the thoracic region every time you _____(inhale or exhale).

Answer: inhale
Type: SA   Page Ref: 480
Topic: LYMPH

5) The lymph from the abdominal digestive organs flows into the _____ trunk.

Answer: intestinal
Type: SA   Page Ref: 481
Topic: LYMPHATIC VESSELS

6) Lymph from the right upper extremity flows through the _____ trunk, then into the _____ duct.

Answer: right subclavian, right lymphatic
Type: SA   Page Ref: 482
Topic: LYMPHATIC VESSELS

7) The _____ duct flows into the left subclavian vein near the junction of the left internal jugular vein and the left _____ vein.

Answer: thoracic, subclavian
Type: SA   Page Ref: 481
Topic: LYMPHATIC VESSELS

8)   Afferent lymphatic vessels have valves that open _____ (toward or away from) a lymph node so that they direct the flow of lymph _____ (into or out of) the node.

Answer: toward, into
Type: SA   Page Ref: 484
Topic: LYMPHATIC TISSUES

9)   The lymph nodes of the abdomen are divided into two groups: _____ and _____.

Answer: parietal, visceral
Type: SA   Page Ref: 491
Topic: LYMPHATIC TISSUES

10) The three types of tonsils, named for their locations, are the pharyngeal, lingual, and _____.

Answer: palatine
Type: SA   Page Ref: 486
Topic: LYMPHATIC TISSUES

11) Lymphatic nodules scattered in the lining of the respiratory, urinary, reproductive, and gastrointestinal tracts are referred to as _____.

Answer: mucosa-associated lymphoid tissue or MALT
Type: SA   Page Ref: 486
Topic: LYMPHATIC TISSUES

12) Extensions of the capsule of the thymus gland called _____ divide the lobes of the gland into lobules.

Answer: trabeculae
Type: SA   Page Ref: 483
Topic: LYMPHATIC TISSUES

MATCHING.   Choose the item in Column 2 that best matches each item in Column 1.

Match the vessels of the lymphatic system in Column 1 with their descriptions in Column 2.

1)   Column 1: lymphatic capillaries
     Column 2: closed-ended tubes

     Answer: closed-ended tubes
     Type: MA   Page Ref: 478
     Topic: LYMPHATIC VESSELS

2)   Column 1: lymphatic vessels
     Column 2: similar in structure to veins

     Answer: similar in structure to veins
     Type: MA   Page Ref: 478
     Topic: LYMPHATIC VESSELS

3) Column 1: afferent lymphatic vessels
   Column 2: carry lymph only into
              lymph nodes
   Foil: carry lymph only away
              from lymph nodes

   Answer: carry lymph only into lymph nodes
   Type: MA   Page Ref: 484
   Topic: LYMPHATIC VESSELS

4) Column 1: thoracic duct
   Column 2: main collecting duct

   Answer: main collecting duct
   Type: MA   Page Ref: 481
   Topic: LYMPHATIC VESSELS

5) Column 1: right lymphatic duct
   Column 2: receives lymph from upper
              right portion of body

   Answer: receives lymph from upper right portion of body
   Type: MA   Page Ref: 481
   Topic: LYMPHATIC VESSELS

6) Column 1: left subclavian trunk
   Column 2: flows into thoracic duct
   Foil: receives lymph from
              thoracic duct

   Answer: flows into thoracic duct
   Type: MA   Page Ref: 481
   Topic: LYMPHATIC VESSELS

7) Column 1: cysterna chyli
   Column 2: inferior portion of thoracic
              duct

   Answer: inferior portion of thoracic duct
   Type: MA   Page Ref: 481
   Topic: LYMPHATIC VESSELS

Match the types of lymphatic tissue in Column 1 with their descriptions in Column 2.
8) Column 1: lymphatic nodule
   Column 2: nonencapsulated oval
              structure located, for
              example, in the appendix

   Answer: nonencapsulated oval structure located, for example, in the appendix
   Type: MA   Page Ref: 486
   Topic: LYMPHATIC TISSUES

9)   Column 1: lymph node
      Column 2: encapsulated oval structure;
              filters lymph

      Answer: encapsulated oval structure; filters lymph
      Type: MA   Page Ref: 484
      Topic: LYMPHATIC TISSUES

10)   Column 1: bone marrow
      Column 2: one of the primary
              lymphatic organs, as is the
              thymus gland

      Answer: one of the primary lymphatic organs, as is the thymus gland
      Type: MA   Page Ref: 482
      Topic: LYMPHATIC TISSUES

11)   Column 1: thymus gland
      Column 2: usually bilobed, located in
              the mediastinum

      Answer: usually bilobed, located in the mediastinum
      Type: MA   Page Ref: 483
      Topic: LYMPHATIC TISSUES

12)   Column 1: spleen
      Column 2: contains red pulp and
              white pulp, does *not* filter
              lymph

      Answer: contains red pulp and white pulp, does *not* filter lymph
      Type: MA   Page Ref: 484
      Topic: LYMPHATIC TISSUES

13)   Column 1: tonsils
      Column 2: special groups of lymphatic
              nodules in the pharyngeal
              and oral organs

      Answer: special groups of lymphatic nodules in the pharyngeal and oral organs
      Type: MA   Page Ref: 486
      Topic: LYMPHATIC TISSUES

Match the lymph nodes in Column 1 with their locations in Column 2.
14)   Column 1: subclavicular
      Column 2: posterior and superior to
              pectoralis minor muscle

      Answer: posterior and superior to pectoralis minor muscle
      Type: MA   Page Ref: 489
      Topic: LYMPHATIC TISSUES

15) Column 1: supratrochlear
Column 2: superior to the medial
epicondyle of the humerus

Answer: superior to the medial epicondyle of the humerus
Type: MA   Page Ref: 489
Topic: LYMPHATIC TISSUES

16) Column 1: superficial inguinal
Column 2: parallel to saphenous vein

Answer: parallel to saphenous vein
Type: MA   Page Ref: 490
Topic: LYMPHATIC TISSUES

17) Column 1: deep inguinal
Column 2: medial to femoral vein

Answer: medial to femoral vein
Type: MA   Page Ref: 490
Topic: LYMPHATIC TISSUES

18) Column 1: internal iliac
Column 2: near internal iliac artery

Answer: near internal iliac artery
Type: MA   Page Ref: 491
Topic: LYMPHATIC TISSUES

19) Column 1: gastric
Column 2: along greater and lesser
curvatures of stomach

Answer: along greater and lesser curvatures of stomach
Type: MA   Page Ref: 491
Topic: LYMPHATIC TISSUES

20) Column 1: sternal
Column 2: alongside internal thoracic
artery

Answer: alongside internal thoracic artery
Type: MA   Page Ref: 494
Topic: LYMPHATIC TISSUES

21) Column 1: bronchopulmonary
Column 2: in the hilus of each lung

Answer: in the hilus of each lung
Type: MA   Page Ref: 494
Topic: LYMPHATIC TISSUES

22)   Column 1: phrenic
      Column 2: on the thoracic surface of
            the diaphragm

      Answer: on the thoracic surface of the diaphragm
      Type: MA   Page Ref: 494
      Topic: LYMPHATIC TISSUES

CHAPTER 16   Nervous Tissue

MULTIPLE CHOICE. Choose the one alternative that best completes the statement or answers the question.

1)   The peripheral nervous system carries _____ motor impulses from the _____ to the _____.
A) efferent, muscles, CNS   B) efferent, CNS, muscles
C) afferent, CNS, muscles   D) afferent, muscles, CNS

Answer: B
Type: MC   Page Ref: 501
Topic: NERVOUS SYSTEM DIVISIONS

2)   The peripheral nervous system carries _____ sensory impulses from the _____ to the _____.
A) efferent, CNS, sense organs   B) efferent, sense organs, CNS
C) afferent, sense organs, CNS   D) afferent, CNS, sense organs

Answer: C
Type: MC   Page Ref: 501
Topic: NERVOUS SYSTEM DIVISIONS

3)   The peripheral nervous system consists of:
A) somatic nervous system and autonomic nervous system.
B) sensory neurons and motor neurons.
C) cranial nerves and spinal nerves.
D) all of the above.

Answer: D
Type: MC   Page Ref: 501
Topic: NERVOUS SYSTEM DIVISIONS

4)   A motor unit in skeletal muscle is activated by a/an _____.
A) sympathetic motor neuron   B) parasympathetic motor neuron
C) somatic motor neuron   D) afferent motor neuron

Answer: C
Type: MC   Page Ref: 501
Topic: NERVOUS SYSTEM DIVISIONS

5)   A main difference between neurons and neuroglia is:
A) neuroglia are found only in the CNS.
B) mature neurons do not normally divide; neuroglia do.
C) neurons are more numerous than neuroglia.
D) neurons are generally smaller than neuroglia.

Answer: B
Type: MC   Page Ref: 501
Topic: HISTOLOGY OF THE NERVOUS SYSTEM: Neuroglia

6) Which of the following statements is *not* true regarding oligodendrocytes?
A) They are smaller than astrocytes.
B) They form myelin sheaths around CNS axons.
C) They line the fluid-filled ventricles of the brain.
D) They are the most common glia cells of the CNS.

Answer: C
Type: MC   Page Ref: 502
Topic: HISTOLOGY OF THE NERVOUS SYSTEM: Neuroglia

7) The following are all functions of astrocytes *except*:
A) they form the blood-brain barrier.
B) they synthesize neurotransmitters.
C) they participate in brain development.
D) they help maintain proper K^+ balance.

Answer: B
Type: MC   Page Ref: 502
Topic: HISTOLOGY OF THE NERVOUS SYSTEM: Neuroglia

8) A synapse is a junction between:
A) two neurons.   B) a neuron and a muscle cell.
C) a neuron and a glandular cell.   D) all of the above.

Answer: D
Type: MC   Page Ref: 503
Topic: HISTOLOGY OF THE NERVOUS SYSTEM: Neurons

9) Most nerves are surrounded by three connective tissue coats: _____ surrounds the nerve, _____ surrounds the fascicles, _____ surrounds individual axons.
A) epineurium, perineurium, endoneurium   B) epineurium, endoneurium, perineurium
C) perineurium, endoneurium, epineurium   D) perineurium, epineruium, endoneurium

Answer: A
Type: MC   Page Ref: 507
Topic: HISTOLOGY OF THE NERVOUS SYSTEM: Neurons

10) The axon hillock is the part of a neuron that:
A) passes nerve impulses to another cell.
B) connects the axon to the neuron cell body.
C) gives rise to axon collaterals.
D) stores neurotransmitters in vesicles.

Answer: B
Type: MC   Page Ref: 503
Topic: HISTOLOGY OF THE NERVOUS SYSTEM: Neurons

11) Dendrites are processes of a neuron that:
A) carry nerve impulses away from the neuron cell body.
B) are usually myelinated.
C) are usually longer than axons.
D) none of the above.

Answer: D
Type: MC   Page Ref: 503
Topic: HISTOLOGY OF THE NERVOUS SYSTEM: Neurons

12) When compared to unmyelinated axons, myelinated axons:
1. are electrically insulated
2. are grey in color
3. have a faster speed of impulse conduction
4. are more numerous

A) 1, 2, 3, 4   B) 1, 3, 4   C) 2, 3, 4   D) 1, 2, 4

Answer: B
Type: MC   Page Ref: 503, 510
Topic: HISTOLOGY OF THE NERVOUS SYSTEM: Neurons

13) A nerve is a structure that may contain:
1. axons
2. dendrites
3. neuron cell bodies
4. connective tissue
5. myelin
6. oligodendrocytes

A) 1, 2, 4, 5   B) 1, 3, 5, 6   C) 1, 4, 5   D) 2, 3, 4

Answer: A
Type: MC   Page Ref: 507
Topic: HISTOLOGY OF THE NERVOUS SYSTEM: Neurons

14) White matter includes:
A) ganglia.
B) a thin outer layer of most of the brain.
C) nuclei in the brain.
D) tracts in the spinal cord.

Answer: D
Type: MC   Page Ref: 510
Topic: GRAY MATTER AND WHITE MATTER

15) Which of the following pairs of terms is most closely matched?
A) gray matter, myelinated axons of the PNS
B) white matter, neuron cell bodies in a ganglion
C) gray matter, neurolemmocytes in the PNS
D) white matter, tracts in the CNS

Answer: D
Type: MC   Page Ref: 510
Topic: GRAY MATTER AND WHITE MATTER

16) At a synapse between the axon of one cell and the dendrite of another cell, the axon would always be part of a _____ neuron; the dendrite would always be part of a _____ neuron.
A) presynaptic, postsynaptic   B) visceral, somatic
C) postsynaptic, presynaptic   D) somatic, visceral

Answer: A
Type: MC   Page Ref: 510
Topic: NEURONAL CIRCUITS

17) Diverging circuits allow for transmission of nerve impulses from _____ neuron(s) to _____ neuron(s).
   A) several, one
   B) one, several
   C) unipolar, multipolar
   D) multipolar, unipolar

   Answer: B
   Type: MC   Page Ref: 510
   Topic: NEURONAL CIRCUITS

18) Regeneration of nerve fibers in the CNS is highly unlikely because:
   A) there are no neurolemmas.
   B) neuroglia inhibit axon regeneration.
   C) scar tissue rapidly forms a physical barrier.
   D) all of the above.

   Answer: D
   Type: MC   Page Ref: 506
   Topic: HISTOLOGY OF THE NERVOUS SYSTEM: Neurons

TRUE/FALSE. Write 'T' if the statement is true and 'F' if the statement is false.

1) The efferent portion of the autonomic nervous system consists of sympathetic and parasympathetic divisions.

   Answer: TRUE
   Type: TF   Page Ref: 501
   Topic: NERVOUS SYSTEM DIVISIONS

2) Neurons that originate in the CNS and that function to carry information to smooth, skeletal, or cardiac muscle are efferent neurons.

   Answer: TRUE
   Type: TF   Page Ref: 501
   Topic: NERVOUS SYSTEM DIVISIONS

3) The afferent portion of the somatic nervous system carries sensory information from the eyes and ears to the CNS.

   Answer: TRUE
   Type: TF   Page Ref: 501
   Topic: NERVOUS SYSTEM DIVISIONS

4) The afferent portion of the autonomic nervous system carries sensory information from the viscera to the CNS.

   Answer: TRUE
   Type: TF   Page Ref: 501
   Topic: NERVOUS SYSTEM DIVISIONS

5) A significant difference between neurons and neuroglia is that mature neuroglia can divide; mature neurons cannot.

   Answer: TRUE
   Type: TF   Page Ref: 501
   Topic: HISTOLOGY OF THE NERVOUS SYSTEM: Neuroglia

6) The neuroglia cells that are derived from monocytes are called astrocytes.

Answer: FALSE
Type: TF   Page Ref: 502
Topic: HISTOLOGY OF THE NERVOUS SYSTEM: Neuroglia

7) Dendrites conduct impulses toward the neuron cell body; the axon conducts impulses away from the neuron cell body.

Answer: TRUE
Type: TF   Page Ref: 503
Topic: HISTOLOGY OF THE NERVOUS SYSTEM: Neurons

8) The portion of the plasma membrane of a neuron that surrounds the axoplasm is called the neurolemma.

Answer: FALSE
Type: TF   Page Ref: 503
Topic: HISTOLOGY OF THE NERVOUS SYSTEM: Neurons

9) The name given to rough endoplasmic reticulum in neurons is lipofuscin granules.

Answer: FALSE
Type: TF   Page Ref: 503
Topic: HISTOLOGY OF THE NERVOUS SYSTEM: Neurons

10) There are two types of neuroglia that produce myelin sheaths: oligodendrocytes and neurolemmocytes.

Answer: TRUE
Type: TF   Page Ref: 503
Topic: HISTOLOGY OF THE NERVOUS SYSTEM: Neuroglia

11) A neurolemma is a characteristic of axons located in the peripheral nervous system, but not in the central nervous system.

Answer: TRUE
Type: TF   Page Ref: 506
Topic: HISTOLOGY OF THE NERVOUS SYSTEM: Neurons

12) The number of association neurons in the body is more than the combined number of motor and sensory neurons.

Answer: TRUE
Type: TF   Page Ref: 509
Topic: CLASSIFICATION OF NEURONS

13) Unipolar neurons are always sensory or afferent neurons.

Answer: TRUE
Type: TF   Page Ref: 507
Topic: CLASSIFICATION OF NEURONS

14) Gray matter is only located in the brain.

Answer: FALSE
Type: TF   Page Ref: 510
Topic: GRAY MATTER AND WHITE MATTER

ESSAY. Write your answer in the space provided or on a separate sheet of paper.

1) Arrange the following terms in a chart that exhibits your understanding of the organization of the nervous system: somatic nervous system, central nervous system, sympathetic division, peripheral nervous system, parasympathetic division, autonomic nervous system.

Answer: The following should be in a simple chart, the format of which may vary, but should be clear and concise:
CNS: brain and spinal cord
PNS: (1) SNS
(2) ANS: sympathetic
parasympathetic
All sections are connected by afferent and efferent pathways, except for the ANS sympathetic and parasympathetic pathways, which are efferent only.
Type: ES   Page Ref: 501, 502
Topic: NERVOUS SYSTEM DIVISIONS

2) List and locate six types of neuroglia cells. Briefly describe the structure and give a main function of each.

Answer: This information is summarized in Table 16.1.
Type: ES   Page Ref: 502, 503
Topic: HISTOLOGY OF THE NERVOUS SYSTEM: Neuroglia

3) Using correct anatomical terms, describe the structure of an axon from the point where it is attached to the neuron cell body to where it ends at a synapse.

Answer: The description should include correct use of the following terms: axon hillock, axolemma, axoplasm, initial segment, trigger zone, mitochondria, microtubules, neurofibrils, axon collaterals, axon terminals, synaptic endbulbs, synaptic vesicles, myelin sheath.
Type: ES   Page Ref: 503
Topic: HISTOLOGY OF THE NERVOUS SYSTEM: Neurons

4) Describe the types of neurons according to stuctural classification and give an example of where each type is located in the body.

Answer: Multipolar, bipolar, and unipolar neurons are defined, with examples on pp. 507.
Type: ES   Page Ref: 507
Topic: CLASSIFICATION OF NEURONS

SHORT ANSWER. Write the word or phrase that best completes each statement or answers the question.

1) The two principal divisions of the nervous system are _____ and _____.

Answer: central, peripheral
Type: SA   Page Ref: 501
Topic: NERVOUS SYSTEM DIVISIONS

2)   The efferent portion of the somatic nervous system is _____ (voluntary or involuntary).

Answer: voluntary
Type: SA   Page Ref: 501
Topic: NERVOUS SYSTEM DIVISIONS

3)   Scar tissue in the nervous system is formed by division and growth of _____.

Answer: neuroglia
Type: SA   Page Ref: 502
Topic: HISTOLOGY OF THE NERVOUS SYSTEM: Neuroglia

4)   One type of neuroglia is found as a single layer of epithelial cells that serve as a lining for fluid–filled spaces in the CNS. This type of neuroglia is _____.

Answer: ependymal cells
Type: SA   Page Ref: 502
Topic: HISTOLOGY OF THE NERVOUS SYSTEM: Neuroglia

5)   An oligodendrocyte contributes to the myelin sheath of _____ (one or several) axon(s).

Answer: several
Type: SA   Page Ref: 507
Topic: HISTOLOGY OF THE NERVOUS SYSTEM: Neuroglia

6)   A neurolemmocyte contributes to the myelin sheath of (one or several) axon(s).

Answer: one
Type: SA   Page Ref: 506, 507
Topic: HISTOLOGY OF THE NERVOUS SYSTEM: Neuroglia

7)   A neuron typically consists of three parts: cell body, _____ and _____.

Answer: axon, dendrites
Type: SA   Page Ref: 503
Topic: HISTOLOGY OF THE NERVOUS SYSTEM: Neurons

8)   Axons are arranged in groups called _____ in nerves.

Answer: fascicles
Type: SA   Page Ref: 507
Topic: HISTOLOGY OF THE NERVOUS SYSTEM: Neurons

9)   A bundle of PNS nerve fibers together with their myelin sheath and associated connective tissue is called a/an _____.

Answer: nerve
Type: SA   Page Ref: 507
Topic: HISTOLOGY OF THE NERVOUS SYSTEM: Neurons

10) A bundle of nerve fibers in the CNS is called a/an _____.

Answer: tract
Type: SA   Page Ref: 507
Topic: HISTOLOGY OF THE NERVOUS SYSTEM: Neurons

11) A mass of neuron cell bodies in the PNS is called a/an _____; in the CNS it is called a/an _____.

Answer: ganglion, nucleus
Type: SA   Page Ref: 507
Topic: HISTOLOGY OF THE NERVOUS SYSTEM: Neurons

12) A neuron having one axon and one dendrite, such as those located in the retina or inner ear, are classified as _____ neurons.

Answer: bipolar
Type: SA   Page Ref: 507
Topic: CLASSIFICATION OF NEURONS

MATCHING.   Choose the item in Column 2 that best matches each item in Column 1.

Match the neuroglia in Column 1 with their descriptions in Column 2.
1) Column 1: oligodendrocytes
   Column 2: most common type in CNS,
             form myelin sheath in CNS

   Answer: most common type in CNS, form myelin sheath in CNS
   Type: MA   Page Ref: 502
   Topic: HISTOLOGY OF THE NERVOUS SYSTEM: Neuroglia

2) Column 1: astrocytes
   Column 2: help form the blood-brain
             barrier

   Answer: help form the blood-brain barrier
   Type: MA   Page Ref: 502
   Topic: HISTOLOGY OF THE NERVOUS SYSTEM: Neuroglia

3) Column 1: microglia
   Column 2: phagocytes of CNS
   Foil: phagocytes of PNS

   Answer: phagocytes of CNS
   Type: MA   Page Ref: 502
   Topic: HISTOLOGY OF THE NERVOUS SYSTEM: Neuroglia

4) Column 1: ependymal cells
   Column 2: lining of brain ventricles

   Answer: lining of brain ventricles
   Type: MA   Page Ref: 502
   Topic: HISTOLOGY OF THE NERVOUS SYSTEM: Neuroglia

5) Column 1: neurolemmocytes
   Column 2: form myelin sheaths of PNS

   Answer: form myelin sheaths of PNS
   Type: MA   Page Ref: 503
   Topic: HISTOLOGY OF THE NERVOUS SYSTEM: Neuroglia

6)   Column 1: satellite cells
        Column 2: surround neurons in PNS
                ganglia

        Answer: surround neurons in PNS ganglia
        Type: MA   Page Ref: 503
        Topic: HISTOLOGY OF THE NERVOUS SYSTEM: Neuroglia

Match the features of a neuron in Column 1 with their descriptions in Column 2.
7)   Column 1: neurofibrils
        Column 2: form(s) part of the
                cytoskeleton

        Answer: form(s) part of the cytoskeleton
        Type: MA   Page Ref: 503
        Topic: HISTOLOGY OF THE NERVOUS SYSTEM: Neurons

8)   Column 1: chromotophilic substance
        Column 2: rough endoplasmis
                reticulum

        Answer: rough endoplasmis reticulum
        Type: MA   Page Ref: 503
        Topic: HISTOLOGY OF THE NERVOUS SYSTEM: Neurons

9)   Column 1: lipofuscin
        Column 2: accumulates as yellow–
                brown granules as the cell
                ages

        Answer: accumulates as yellow–brown granules as the cell ages
        Type: MA   Page Ref: 503
        Topic: HISTOLOGY OF THE NERVOUS SYSTEM: Neurons

10)   Column 1: dendrites
        Column 2: usually short and branched

        Answer: usually short and branched
        Type: MA   Page Ref: 503
        Topic: HISTOLOGY OF THE NERVOUS SYSTEM: Neurons

11)   Column 1: axom
        Column 2: usually myelinated

        Answer: usually myelinated
        Type: MA   Page Ref: 503
        Topic: HISTOLOGY OF THE NERVOUS SYSTEM: Neurons

12)   Column 1: neurotransmitter molecules
        Column 2: stores in synaptic vesicles

        Answer: stores in synaptic vesicles
        Type: MA   Page Ref: 503
        Topic: HISTOLOGY OF THE NERVOUS SYSTEM: Neurons

13) Column 1: axon terminals
Column 2: fine processes at the ends
of axons
Foil: join axon to cell body

Answer: fine processes at the ends of axons
Type: MA   Page Ref: 503
Topic: HISTOLOGY OF THE NERVOUS SYSTEM: Neurons

14) Column 1: nerve fiber
Column 2: may refer to either axons or
dendrites

Answer: may refer to either axons or dendrites
Type: MA   Page Ref: 507
Topic: HISTOLOGY OF THE NERVOUS SYSTEM: Neurons

CHAPTER 17   The Spinal Cord and the Spinal Nerves

MULTIPLE CHOICE.   Choose the one alternative that best completes the statement or answers the question.

1)   The meninges are layers of connective tissue:
      A) that surround the brain and spinal cord.
      B) that cover the spinal nerves up to where they exit through the intervertebral foramina.
      C) that are called, in order from external to infernal: dura mater, arachnoid, and pia mater.
      D) all of the above.

      Answer: D
      Type: MC   Page Ref: 516
      Topic: SPINAL CORD ANATOMY

2)   The arachnoid layer of the meninges lies between two fluids: _____ fluid on its outer surface and _____ fluid on its inner surface.
      A) interstitial, blood   B) cerebrospinal, interstitial
      C) blood, cerebrospinal   D) interstitial, cerebrospinal

      Answer: D
      Type: MC   Page Ref: 516, 517
      Topic: SPINAL CORD ANATOMY

3)   The conus medullaris is:
      A) the junction between the medulla and the spinal cord.
      B) the taper of the spinal cord inferior to the lumbar enlargement.
      C) the inner portion of the spinal cord, seen in cross section.
      D) the attachment of a spinal nerve to the spinal cord.

      Answer: B
      Type: MC   Page Ref: 517
      Topic: SPINAL CORD ANATOMY

4)   The inferior extension of the pia mater that anchors the spinal cord to the coccyx is called the _____.
      A) cauda equina   B) filum terminale
      C) denticulate ligament   D) conus medullaris

      Answer: B
      Type: MC   Page Ref: 517
      Topic: SPINAL CORD ANATOMY

5)   The white matter of the spinal cord:
      A) contains sensory and motor, or ascending and descending, tracts.
      B) is surrounded by gray matter.
      C) is subdivided into regions called horns.
      D) all of the above.

      Answer: A
      Type: MC   Page Ref: 520
      Topic: SPINAL CORD ANATOMY

6)   The gray matter of the spinal cord:
     A) is subdivided into regions called horns.
     B) is surrounded by white matter.
     C) contains neuroglia, neuron cell bodies, and unmyelinated axons and dendrites.
     D) all of the above.

     Answer: D
     Type: MC   Page Ref: 520
     Topic: SPINAL CORD ANATOMY

7)   The central canal of the spinal cord is located in the:
     A) anterior white commissure.   B) anterior gray horn.
     C) posterior white column.   D) gray commissure.

     Answer: D
     Type: MC   Page Ref: 517
     Topic: SPINAL CORD ANATOMY

8)   Lateral gray horns are present in all segments of the _____ region of the spinal cord.
     A) cervical   B) thoracic   C) lumbar   D) sacral

     Answer: B
     Type: MC   Page Ref: 520
     Topic: SPINAL CORD ANATOMY

9)   The senses of pain and temperature are conveyed to the brain by:
     A) pyramidal tracts.   B) extrapyramidal tracts.
     C) spinothalamic tracts.   D) posterior column tracts.

     Answer: C
     Type: MC   Page Ref: 521
     Topic: SPINAL CORD FUNCTIONS

10) Motor impulses that help maintain muscle tone and posture are conveyed from the
     brain through the spinal cord by the:
     A) direct pathways.   B) indirect pathways.
     C) spinothalamic tracts.   D) posterior column tracts.

     Answer: B
     Type: MC   Page Ref: 522
     Topic: SPINAL CORD FUNCTIONS

11) A monosynaptic reflex arc is a pathway that:
     A) contains only one neuron.
     B) consists of a sensory neuron, an association neuron, and a motor neuron.
     C) consists of a sensory neuron and an association neuron.
     D) none of the above.

     Answer: D
     Type: MC   Page Ref: 522
     Topic: SPINAL CORD FUNCTIONS

12) A polysynaptic reflex arc involves, in order, the following components:
1. sensory neuron
2. motor neuron
3. receptor
4. one or more association neurons
5. effector

A) 3, 1, 4, 2, 5   B) 3, 4, 1, 2, 5   C) 3, 1, 2, 5   D) 3, 2, 5

Answer: A
Type: MC   Page Ref: 522
Topic: SPINAL CORD FUNCTIONS

13) A pathway called a reflex arc terminates at an effector, which could be:

A) skeletal muscles.   B) smooth or cardiac muscle.
C) glands.   D) all of the above.

Answer: D
Type: MC   Page Ref: 523
Topic: SPINAL CORD FUNCTIONS

14) The roots of which spinal nerves make up the cauda equina?
1. thoracic
2. lumbar
3. sacral
4. coccygeal

A) 1, 2, 3, 4   B) 2, 3, 4   C) 3, 4   D) 4 only

Answer: B
Type: MC   Page Ref: 523
Topic: SPINAL NERVES

15) Which of the following is *not* true for spinal nerves?
A) Each spinal nerve has two roots; an anterior motor root and a posterior sensory root.
B) The outer covering of spinal nerve roots is dura mater.
C) The outer covering of spinal nerves is epineurium.
D) A spinal nerve exits the vertebral column via the vertebral foramen.

Answer: D
Type: MC   Page Ref: 517, 523
Topic: SPINAL NERVES

16) Which of the following is *not* true regarding the sciatic nerve? It:
A) arises from the lumbar plexus.
B) splits at the level of the knee into tibial and common peroneal nerves.
C) is the largest nerve in the body.
D) innervates the hamstring muscles.

Answer: A
Type: MC   Page Ref: 538
Topic: SPINAL NERVES

17) The main plexuses formed by the ventral rami of spinal nerves are:
   A) thoracic, lumbar, sacral, and coccygeal   B) cervical, brachial, lumbar, and sacral.
   C) cervical, lumbar, sacral, and inguinal.   D) cervical, lumbar, sacral, and coccygeal.

Answer: B
Type: MC   Page Ref: 524
Topic: SPINAL NERVES

18) Spinal nerves T2–T12 differ from all other spinal nerves in that:
   A) they do not branch to form rami.
   B) they are autonomic nerves.
   C) the ventral rami do not contribute to a plexus.
   D) the dorsal rami form a plexus.

Answer: C
Type: MC   Page Ref: 524
Topic: SPINAL NERVES

19) Intercostal muscles and overlying skin are innervated by:
   A) nerves from the cervical plexus   B) ventral rami of thoracic spinal nerves
   C) intercostal nerves   D) either B) or C) is a correct response

Answer: D
Type: MC   Page Ref: 525
Topic: SPINAL NERVES

TRUE/FALSE. Write 'T' if the statement is true and 'F' if the statement is false.

1) The epidural space found between the dura mater and the wall of the vertebral canal contains a protective cushion of fat and connective tissue.

Answer: TRUE
Type: TF   Page Ref: 516
Topic: SPINAL CORD ANATOMY

2) The thin transparent spinal meninx called pia mater separates the spinal cord from the surrounding cerebrospinal fluid.

Answer: TRUE
Type: TF   Page Ref: 516
Topic: SPINAL CORD ANATOMY

3) Lateral extensions of the pia mater, called denticulate ligaments, suspend the spinal cord in the middle of the dural sheath, providing protection against shock and sudden displacement.

Answer: TRUE
Type: TF   Page Ref: 517
Topic: SPINAL CORD ANATOMY

4) The anterior median fissure of the spinal cord is shallower and narrower than the posterior median sulcus.

Answer: FALSE
Type: TF   Page Ref: 517
Topic: SPINAL CORD ANATOMY

5) Motor neurons that supply smooth muscle, cardiac muscle, or glands via the autonomic nervous system can be found in the lateral gray horns of the spinal cord.
Answer: TRUE
Type: TF   Page Ref: 520
Topic: SPINAL CORD FUNCTIONS

6) Plexuses are formed by the ventral rami of all cervical, lumbar, and sacral nerves, and the ventral rami of only the first thoracic nerves.
Answer: TRUE
Type: TF   Page Ref: 524
Topic: SPINAL NERVES

7) The femoral nerve is a branch of the sciatic nerve.
Answer: FALSE
Type: TF   Page Ref: 534
Topic: SPINAL NERVES

8) Complete paralysis of the diaphragm occurs if the spinal cord is severed just below the fifth cervical nerve.
Answer: FALSE
Type: TF   Page Ref: 526
Topic: SPINAL NERVES

ESSAY. Write your answer in the space provided or on a separate sheet of paper.

1) Describe, from outermost to innermost, the structures, tissues, and fluids that provide protection for the spinal cord.
Answer: 1. vertebral canal of the vertebral column, 2. vertebral ligaments, 3. epidural space with adipose tissue, 4. dura mater of dense irregular connective tissue, 5. subdural space with interstitial fluid, 6. arachnoid layer of collagen and elastic fibers, 7. subarachnoid space with CSF, 8. pia mater of thin fibrous connective tissue, 9. denticulate ligaments, extensions of pia mater, attach cord to dura mater.
Type: ES   Page Ref: 516, 517
Topic: SPINAL CORD ANATOMY

2) Sketch and label a cross section of a spinal cord segment showing the attachments of a spinal nerve and the features of the cord.
Answer: See Fig. 17.3.
Type: ES   Page Ref: 516, 517
Topic: SPINAL CORD ANATOMY

3) Describe the anatomy of a spinal nerve from the point where it is attached to the spinal cord to where it divides into branches. Name the roots and branches and describe the regions served by each branch.

Answer: The nerve is attached to the cord by the anterior root and the posterior root (which contains the sensory cell bodies in a posterior root ganglion). The two roots merge to form a single spinal nerve, which passes through an intervertebral foramen. The spinal nerve then divides into:
1. the anterior (ventral) ramus, which serves the upper and lower limbs and lateral and anterior trunk,
2. the posterior (dorsal) ramus, which serves the posterior region of the trunk,
3. the meningeal branch, which reenters the spinal canal to serve the meninges, vertebrae, vertebral ligaments, and blood vessels of the spinal cord, and
4. rami communicantes, which contain components of autonomic nerve pathways.
Type: ES   Page Ref: 524
Topic: SPINAL NERVES

SHORT ANSWER. Write the word or phrase that best completes each statement or answers the question.

1) The adult spinal cord extends from the _____ of the brain inferiorly to the level of the _____ lumbar vertebra.

Answer: medulla oblongata, second
Type: SA   Page Ref: 517
Topic: SPINAL CORD ANATOMY

2) Cerebrospinal fluid is found between two layers of the meninges, in a space called the _____.

Answer: subarachnoid space
Type: SA   Page Ref: 517
Topic: SPINAL CORD ANATOMY

3) The pia mater of the spinal cord lies next to the _____ (gray or white) matter of the cord.

Answer: white
Type: SA   Page Ref: 516, 517
Topic: SPINAL CORD ANATOMY

4) The two main routes for sensory information in the spinal cord are _____ tracts and _____ tracts.

Answer: spinothalamic, posterior column
Type: SA   Page Ref: 521
Topic: SPINAL CORD FUNCTIONS

5) The two main routes for motor information in the spinal cord are _____ pathways and _____ pathways.

Answer: direct, indirect
Type: SA   Page Ref: 522
Topic: SPINAL CORD FUNCTIONS

6) Sensory impulses arrive at the spinal cord via the _____ root of a spinal nerve.

Answer: posterior (dorsal)
Type: SA   Page Ref: 523
Topic: SPINAL CORD FUNCTIONS

7) Motor neurons that supply skeletal muscles have cell bodies in _____ (anterior, posterior, or lateral) gray horns of the spinal cord, and their axons exit the cord via a/an _____ (anterior or posterior) root.

Answer: anterior, anterior
Type: SA   Page Ref: 520, 522
Topic: SPINAL CORD FUNCTIONS

8) There are _____ pairs of spinal nerves, consisting of the following groups: _____ cervical, _____ thoracic, _____ lumbar, _____ sacral, _____ coccygeal.

Answer: 31, 8, 12, 5, 5, 1
Type: SA   Page Ref: 523
Topic: SPINAL NERVES

9) The branches of a spinal nerve are _____, _____, _____, and _____.

Answer: posterior ramus, anterior ramus, meningeal branch, rami communicantes
Type: SA   Page Ref: 524
Topic: SPINAL NERVES

10) The _____ plexus serves the skin and muscles of the head, neck, and superior part of the shoulders and chest.

Answer: cervical
Type: SA   Page Ref: 526
Topic: SPINAL NERVES

11) A _____ is an area of skin that provides sensory input to one pair of spinal nerves.

Answer: dermatome
Type: SA   Page Ref: 525
Topic: SPINAL NERVES

MATCHING.   Choose the item in Column 2 that best matches each item in Column 1.

Match the spinal meninges or spaces in Column 1 with their descriptions in Column 2.
1) Column 1: epidural space
   Column 2: contains a layer of adipose
                      tissue
   Foil: between the dura mater
                      and arachnoid

   Answer: contains a layer of adipose tissue
   Type: MA   Page Ref: 516
   Topic: SPINAL CORD ANATOMY

2) Column 1: subarachnoid space
Column 2: between arachnoid and pia
         mater
Foil: between pia mater and
         spinal cord

Answer: between arachnoid and pia mater
Type: MA   Page Ref: 517
Topic: SPINAL CORD ANATOMY

3) Column 1: pia mater
Column 2: contains blood vessels that
         supply oxygen and
         nutrients to the spinal cord

Answer: contains blood vessels that supply oxygen and nutrients to the spinal cord
Type: MA   Page Ref: 516
Topic: SPINAL CORD ANATOMY

4) Column 1: arachnoid
Column 2: consists of collagen and
         elastic fiber network; has
         no blood vessels

Answer: consists of collagen and elastic fiber network; has no blood vessels
Type: MA   Page Ref: 516
Topic: SPINAL CORD ANATOMY

5) Column 1: dura mater
Column 2: meninx closest to the bone
         of the vertebral column

Answer: meninx closest to the bone of the vertebral column
Type: MA   Page Ref: 516
Topic: SPINAL CORD ANATOMY

Match the nerves in Column 1 with the distribution in Column 2.
6) Column 1: axillary nerve
Column 2: deltoid muscle

Answer: deltoid muscle
Type: MA   Page Ref: 529
Topic: SPINAL NERVES

7) Column 1: radial nerve
Column 2: extensors of the wrist

Answer: extensors of the wrist
Type: MA   Page Ref: 529
Topic: SPINAL NERVES

8) Column 1: median nerve
Column 2: flexors of the wrist

Answer: flexors of the wrist
Type: MA   Page Ref: 529
Topic: SPINAL NERVES

9) Column 1: phrenic
   Column 2: diaphragm

   Answer: diaphragm
   Type: MA   Page Ref: 526
   Topic: SPINAL NERVES

10) Column 1: femoral
    Column 2: quadriceps femoris muscle

    Answer: quadriceps femoris muscle
    Type: MA   Page Ref: 534
    Topic: SPINAL NERVES

11) Column 1: obturator
    Column 2: adductor muscles of thigh

    Answer: adductor muscles of thigh
    Type: MA   Page Ref: 534
    Topic: SPINAL NERVES

12) Column 1: tibial
    Column 2: gastrocnemius and soleus
           muscles

    Answer: gastrocnemius and soleus muscles
    Type: MA   Page Ref: 538
    Topic: SPINAL NERVES

13) Column 1: common peroneal
    Column 2: tibialis anterior and
           peroneal muscles

    Answer: tibialis anterior and peroneal muscles
    Type: MA   Page Ref: 539
    Topic: SPINAL NERVES

Match the nerves in Column 1 with the plexuses in Column 2.
14) Column 1: long thoracic
    Column 2: brachial

    Answer: brachial
    Type: MA   Page Ref: 529
    Topic: SPINAL NERVES

15) Column 1: great auricular
    Column 2: cervical

    Answer: cervical
    Type: MA   Page Ref: 526
    Topic: SPINAL NERVES

16) Column 1: femoral
    Column 2: lumbar

    Answer: lumbar
    Type: MA   Page Ref: 534
    Topic: SPINAL NERVES

17)   Column 1: phrenic
        Column 2: cervical

        Answer: cervical
        Type: MA   *Page Ref: 526*
        Topic: SPINAL NERVES

18)   Column 1: superior gluteal
        Column 2: sacral

        Answer: sacral
        Type: MA   *Page Ref: 538*
        Topic: SPINAL NERVES

19)   Column 1: ulnar
        Column 2: brachial

        Answer: brachial
        Type: MA   *Page Ref: 529*
        Topic: SPINAL NERVES

20)   Column 1: pudendal
        Column 2: sacral

        Answer: sacral
        Type: MA   *Page Ref: 538*
        Topic: SPINAL NERVES

21)   Column 1: iliohypogastric
        Column 2: lumbar

        Answer: lumbar
        Type: MA   *Page Ref: 534*
        Topic: SPINAL NERVES

CHAPTER 18   The Brain and the Cranial Nerves

MULTIPLE CHOICE. Choose the one alternative that best completes the statement or answers the question.

1)   Choroid plexuses are specialized capillaries in the CNS that:
   1. are covered by ependymal cells
   2. are located in the walls of the ventricles
   3. are located in the dural venous sinuses
   4. are the sites for reabsorption of CSF
   5. are derived from the arachnoid layer of the meninges

   A) 3, 4, 5   B) 1, 2, 4, 5   C) 1, 2   D) 2, 3, 4

   Answer: C
   Type: MC   Page Ref: 547
   Topic: CEREBROSPINAL FLUID

2)   Which of the following is *not* a feature of the cranial meninges?
   A) dural sinuses   B) falx cerebri
   C) tentorium cerebelli   D) epidural space

   Answer: D
   Type: MC   Page Ref: 546
   Topic: CRANIAL MENINGES

3)   CSF contains all of the following *except*:
   A) erythrocytes.   B) lymphocytes.   C) water.   D) electrolytes.

   Answer: A
   Type: MC   Page Ref: 547
   Topic: CEREBROSPINAL FLUID

4)   The blood–cerebrospinal fluid barrier:
   A) is the same thing as the blood-brain barrier.
   B) is due mainly to tight junctions between endothelial cells of the capillaries throughout the brain tissue.
   C) protects the neurons of the brain from potentially harmful substances in the blood.
   D) is another name for cranial meninges.

   Answer: C
   Type: MC   Page Ref: 547
   Topic: CEREBROSPINAL FLUID

5)   Circumventricular organs in the walls of the third and fourth ventricles lack a blood-brain barrier. This enables them to:
   A) facilitate the circulation of cerebrospinal fluid in the ventricles.
   B) monitor the chemical composition of the blood.
   C) secrete cerebrospinal fluid into the ventricles.
   D) all of the above.

   Answer: B
   Type: MC   Page Ref: 541
   Topic: BLOOD SUPPLY OF THE BRAIN

6) Which of the following is an *incorrect* statement about a feature of the medulla oblongata?
A) The inferior cerebellar peduncles are fiber tracts that connect the olives of the medulla to the cerebellum.
B) The vestibular nuclear complex is found mostly in the medulla.
C) Pyramids contain the main sensory tracts that pass through the brain stem from the spinal cord; they are visible on the dorsal surface of the medulla.
D) Vital reflex centers for control of heart rate, breathing rate, and blood pressure are located in the medulla.

Answer: C
Type: MC   Page Ref: 551
Topic: BRAIN STEM

7) The regions of the brain stem involved in the control of respiration are the:
A) medulla and pons.
B) pons and midbrain.
C) midbrain and medulla.
D) midbrain, pons, and medulla.

Answer: A
Type: MC   Page Ref: 553, 555
Topic: BRAIN STEM

8) The tectum of the midbrain bears four rounded elevations:
A) called substantia nigra.
B) that control subconscious muscle activities.
C) that transmit all sensory and motor information between the upper and lower brain regions.
D) that control reflex movements in response to visual and auditory stimuli.

Answer: D
Type: MC   Page Ref: 555
Topic: BRAIN STEM

9) Which of the following thalamic nuclei is correctly matched with its function?
A) anterior nucleus; emotions and memory
B) lateral geniculate nucleus; taste
C) medial geniculate nucleus; voluntary motor actions
D) ventral posterior nucleus; vision

Answer: A
Type: MC   Page Ref: 558
Topic: DIENCEPHALON

10) Which of the following is *not* a part of the hypothalamus?
A) preoptic region
B) tuberal region
C) supraoptic region
D) intermediate mass

Answer: D
Type: MC   Page Ref: 556
Topic: DIENCEPHALON

11) The infundibulum is the anatomical link between the brain and the pituitary gland, and therefore it is the primary link between the nervous and endocrine systems. Structures contained in the infundibulum consist of which of the following?
A) a tract that transports hormones from the supraoptic nuclei to the posterior pituitary
B) small blood vessels that transport regulating hormones from the median eminence to the anterior pituitary
C) Both A) and B) are correct.
D) Neither A) nor B) is correct.

Answer: C
Type: MC   Page Ref: 559
Topic: DIENCEPHALON

12) A midsagittal section of the brain would pass through which of the following:
A) longitudinal fissure.   B) corpus callosum.
C) falx cerebri.   D) all of the above.

Answer: D
Type: MC   Page Ref: 559
Topic: CEREBRUM

13) The postcentral gyrus is in the _____ lobe of the cortex; it contains the primary _____ area.
A) parietal, somatosensory   B) parietal, motor
C) frontal, somatosensory   D) frontal, motor

Answer: A
Type: MC   Page Ref: 561
Topic: CEREBRUM

14) Sensory information that arrives in the brain from tracts in the spinal cord is relayed by the thalamus along _____ fibers to the cerebral cortex.
A) association   B) commissural
C) projection   D) none of the above

Answer: C
Type: MC   Page Ref: 561
Topic: CEREBRUM

15) The corpus callosum allows information to travel between cerebral hemispheres. Therefore the corpus callosum contains _____ fibers.
A) association   B) commissural
C) projection   D) all of the above

Answer: B
Type: MC   Page Ref: 561
Topic: CEREBRUM

16) Small ridges on the surface of the cerebellar hemispheres are called:
A) gyri.   B) sulci.   C) folia.   D) vermis.

Answer: C
Type: MC   Page Ref: 568
Topic: CEREBELLUM

17) The cerebellar peduncles conduct information into and out of the cerebellum; for example:
A) the superior cerebellar peduncles conduct only sensory information.
B) the inferior cerebellar peduncles contain only motor fibers.
C) the middle cerebellar peduncles contain only afferent (sensory) fibers.
D) none of the above.

Answer: C
Type: MC   Page Ref: 568
Topic: CEREBELLUM

18) Loss of taste in the anterior region of the tongue may be an indication of damage to which cranial nerve?
A) facial VII   B) trigeminal V
C) glossopharyngeal IX   D) vagus X

Answer: A
Type: MC   Page Ref: 574
Topic: CRANIAL NERVES

19) Inability to control eyeball movement may indicate damage to which cranial nerve(s)?
1. trochlear IV
2. optic II
3. oculomotor III
4. trigeminal V
5. abducens VI
A) 1, 2, 3, 4, 5   B) 2, 3, 4, 5   C) 1, 2, 4   D) 1, 3, 5

Answer: D
Type: MC   Page Ref: 571, 574
Topic: CRANIAL NERVES

20) Which of the following pairs of terms is *not* correctly matched?
A) trigeminal V; largest cranial nerve
B) vagus X; arises from both the brain stem and spinal cord
C) trochlear IV; smallest cranial nerve
D) glossopharyngeal IX; passes through the jugular foramen

Answer: B
Type: MC   Page Ref: 575, 576
Topic: CRANIAL NERVES

TRUE/FALSE. Write 'T' if the statement is true and 'F' if the statement is false.

1) The septum pellucidum is the lining of the third ventricle.
Answer: FALSE
Type: TF   Page Ref: 547
Topic: CEREBROSPINAL FLUID

2) Small openings in the roof of the third ventricle allow passage of CSF into the subarachnoid space.
Answer: FALSE
Type: TF   Page Ref: 547, 548
Topic: CEREBROSPINAL FLUID

3)   Capillaries are unnecessary in nervous tissue of the CNS due to the existence of cerebrospinal fluid, which fulfills the functions of blood.

Answer: FALSE
Type: TF   Page Ref: 547, 550, 551
Topic: CEREBROSPINAL FLUID

4)   Oxygen, carbon dioxide, glucose, and most anesthetics readily pass through the blood-brain barrier.

Answer: TRUE
Type: TF   Page Ref: 551
Topic: BLOOD SUPPLY OF THE BRAIN

5)   Control centers for coughing, sneezing, vomiting, and swallowing are located in the medulla oblongata.

Answer: TRUE
Type: TF   Page Ref: 553
Topic: BRAIN STEM

6)   The pineal gland is attached to the hypothalamus by the infundibulum.

Answer: FALSE
Type: TF   Page Ref: 556, 558
Topic: DIENCEPHALON

7)   The transverse fissure separates the cerebrum from the cerebellum.

Answer: TRUE
Type: TF   Page Ref: 568
Topic: CEREBRUM

8)   The lateral cerebral sulcus separates the temporal lobe from the parietal lobe.

Answer: FALSE
Type: TF   Page Ref: 561
Topic: CEREBRUM

9)   The caudate nucleus and lenticular nucleus are examples of basal ganglia in the cerebrum.

Answer: TRUE
Type: TF   Page Ref: 561
Topic: CEREBRUM

10) Cranial nerves, like spinal nerves, are all mixed nerves.

Answer: FALSE
Type: TF   Page Ref: 568
Topic: CRANIAL NERVES

ESSAY. Write your answer in the space provided or on a separate sheet of paper.

1) Discuss the location of cerebrospinal fluid (CSF) in relation to the three basic functions of CSF; i.e., how does location facilitate function?

Answer: CSF completely surrounds the brain and spinal cord, in the subarachnoid space, acting as a shock absorber. Its location next to the pia mater in the subarachnoid space at the surface of the brain and spinal cord allow CSF to influence and maintain ionic composition of fluid surrounding the CNS neurons. CSF can also deliver nutrients to and remove wastes from the CNS nervous tissue.

Type: ES   Page Ref: 547
Topic: CEREBROSPINAL FLUID

2) Describe the route traveled by cerebrospinal fluid from production until it is reabsorbed into venous blood.

Answer: Beginning with a choroid plexus in a lateral ventricle, the route should include all ventricles, pores, passageways, spaces, etc., as in Fig. 18.3 and as described on pp. 547.

Type: ES   Page Ref: 547
Topic: CEREBROSPINAL FLUID

3) Describe the blood–brain barrier and explain how it controls the movement of substances between the blood and brain tissue. Give examples of substances that readily cross the BBB.

Answer: The following should be described or explained: tight junctions of endothelial cells, basement membrane, astrocytes, lipid solubility, transporters. Substances that cross the BBB include oxygen, carbon dioxide, glucose, anesthetics, water, alcohol, caffeine, heroin, and nicotine.

Type: ES   Page Ref: 551
Topic: BLOOD SUPPLY OF THE BRAIN

4) List the four main regions of the hypothalamus. Name one nucleus or mass of gray matter located in each region and state its function.

Answer: The four regions and their main features are summarized in the text, pp. 558, 559.

Type: ES   Page Ref: 558, 559
Topic: DIENCEPHALON

5) List, in order from from anterior to posterior, the 12 pairs of cranial nerves. Indicate whether each is sensory, motor, or mixed in function. State a main function for each nerve.

Answer: Refer to Table 18.2.
Type: ES   Page Ref: 579–583
Topic: CRANIAL NERVES

SHORT ANSWER. Write the word or phrase that best completes each statement or answers the question.

1)   The _____ is a passageway through the midbrain. It connects the third and fourth ventricles.

Answer: cerebral aqueduct
Type: SA   Page Ref: 547
Topic: CEREBROSPINAL FLUID

2)   Reabsorption of CSF is accomplished by structures in the dural venous sinuses called _____.

Answer: arachnoid villi
Type: SA   Page Ref: 548
Topic: CEREBROSPINAL FLUID

3)   The _____ space is located between the blood vessels of the brain and their surrounding layer of pia mater.

Answer: perivascular
Type: SA   Page Ref: 551
Topic: BLOOD SUPPLY OF THE BRAIN

4)   The crossing of the motor tracts from left to right, and vice versa, as they pass through the medulla is referred to as the _____ of the _____.

Answer: decussation, pyramids
Type: SA   Page Ref: 551
Topic: BRAIN STEM

5)   The nucleus _____ and nucleus _____ are located on the dorsal aspect of the medulla oblongata.

Answer: gracilis, cuneatus
Type: SA   Page Ref: 552, 552
Topic: BRAIN STEM

6)   Small masses of gray matter that are scattered throughout the brain stem and diencephalon and that influence muscle tone and arousal from sleep are collectively referred to as the _____.

Answer: reticular formation
Type: SA   Page Ref: 555
Topic: BRAIN STEM

7)   Two nuclei in the posterior hypothalamus that serve as relay stations for reflexes related to the sense of smell are the _____.

Answer: mammillary bodies
Type: SA   Page Ref: 558
Topic: DIENCEPHALON

8) Name three centers in the hypothalamus that control food and water intake: _____, _____, and _____ centers.

Answer: feeding, satiety, thirst
Type: SA   Page Ref: 559
Topic: DIENCEPHALON

9) White matter arranged in a treelike pattern in the cerebellum is called _____.

Answer: arbor vitae
Type: SA   Page Ref: 561
Topic: CEREBELLUM

10) The area of the brain called the _____ consists of connected masses of gray matter and is involved in short-term memory formation and responses to emotion.

Answer: limbic system
Type: SA   Page Ref: 563, 564
Topic: LIMBIC SYSTEM

11) A recording of the total activity of cells in the cerebral cortex is called a/an _____.

Answer: electroencephalogram
Type: SA   Page Ref: 567
Topic: CEREBRUM: Functional Areas

12) The three branches of the trigeminal V nerve are _____, _____, and _____.

Answer: ophthalmic, maxillary, mandibular
Type: SA   Page Ref: 571
Topic: CRANIAL NERVES

13) The cranial nerve known as the _____ contains motor fibers and sensory fibers that connect to most organs of the ventral body cavity.

Answer: vagus
Type: SA   Page Ref: 575
Topic: CRANIAL NERVES

14) The cranial nerve that originates from both the brain and the spinal cord is the _____.

Answer: accessory XI
Type: SA   Page Ref: 576
Topic: CRANIAL NERVES

15) All cranial nerves, except for _____ and _____, originate from the brain stem.

Answer: olfactory I, optic II
Type: SA   Page Ref: 571
Topic: CRANIAL NERVES

MATCHING. Choose the item in Column 2 that best matches each item in Column 1.

Match the functional areas of the cerebral cortex in Column 1 with the correct lobe of the cortex in Column 2.

1) Column 1: primary visual area
    Column 2: occipital lobe

    Answer: occipital lobe
    Type: MA   Page Ref: 565
    Topic: CEREBRUM: Functional Areas

2) Column 1: primary auditory area
    Column 2: temporal lobe

    Answer: temporal lobe
    Type: MA   Page Ref: 566
    Topic: CEREBRUM: Functional Areas

3) Column 1: primary somatosensory area
    Column 2: parietal lobe

    Answer: parietal lobe
    Type: MA   Page Ref: 564
    Topic: CEREBRUM: Functional Areas

4) Column 1: Broca's (motor speech) area
    Column 2: frontal lobe

    Answer: frontal lobe
    Type: MA   Page Ref: 566
    Topic: CEREBRUM: Functional Areas

5) Column 1: primary motor area
    Column 2: frontal lobe

    Answer: frontal lobe
    Type: MA   Page Ref: 566
    Topic: CEREBRUM: Functional Areas

Match the specialized regions of the brain in Column 1 with their locations in Column 2.

6) Column 1: cerebral peduncles
    Column 2: midbrain

    Answer: midbrain
    Type: MA   Page Ref: 555
    Topic: BRAIN STEM

7) Column 1: cerebral aqueduct
    Column 2: midbrain

    Answer: midbrain
    Type: MA   Page Ref: 555
    Topic: BRAIN STEM

8)   Column 1: olive
        Column 2: medulla oblongata

        Answer: medulla oblongata
        Type: MA   Page Ref: 551
        Topic: BRAIN STEM

9)   Column 1: tuber cinereum
        Column 2: hypothalamus

        Answer: hypothalamus
        Type: MA   Page Ref: 558
        Topic: DIENCEPHALON

10)   Column 1: corpus striatum
        Column 2: cerebrum

        Answer: cerebrum
        Type: MA   Page Ref: 561
        Topic: CEREBRUM

11)   Column 1: medial and lateral
                  geniculate nuclei
        Column 2: thalamus

        Answer: thalamus
        Type: MA   Page Ref: 558
        Topic: DIENCEPHALON

12)   Column 1: insula
        Column 2: cerebrum

        Answer: cerebrum
        Type: MA   Page Ref: 561
        Topic: CEREBRUM

13)   Column 1: hippocampus
        Column 2: limbic system

        Answer: limbic system
        Type: MA   Page Ref: 563
        Topic: LIMBIC SYSTEM

14)   Column 1: arbor vitae
        Column 2: cerebellum

        Answer: cerebellum
        Type: MA   Page Ref: 568
        Topic: CEREBELLUM

15)   Column 1: pneumotaxic and apneustic
                  areas
        Column 2: pons

        Answer: pons
        Type: MA   Page Ref: 555
        Topic: BRAIN STEM

16) Column 1: cardiovascular center
Column 2: medulla oblongata

Answer: medulla oblongata
Type: MA   Page Ref: 553
Topic: BRAIN STEM

Match the cranial nerves in Column 1 to their functions in Column 2.
17) Column 1: olfactory I
Column 2: conveys nerve impulses
related to smell

Answer: conveys nerve impulses related to smell
Type: MA   Page Ref: 579
Topic: CRANIAL NERVES

18) Column 1: optic II
Column 2: conveys nerve impulses
related to vision

Answer: conveys nerve impulses related to vision
Type: MA   Page Ref: 579
Topic: CRANIAL NERVES

19) Column 1: oculomotor III
Column 2: movement of eyelid and
eyeball, constriction of pupil

Answer: movement of eyelid and eyeball, constriction of pupil
Type: MA   Page Ref: 579
Topic: CRANIAL NERVES

20) Column 1: trochlear IV
Column 2: movement of eyeball via
superior oblique muscle

Answer: movement of eyeball via superior oblique muscle
Type: MA   Page Ref: 580
Topic: CRANIAL NERVES

21) Column 1: trigeminal V
Column 2: conveys sensory impulses
from the facial region and
anterior scalp

Answer: conveys sensory impulses from the facial region and anterior scalp
Type: MA   Page Ref: 580
Topic: CRANIAL NERVES

22) Column 1: abducens VI
Column 2: turns eyeball laterally via
lateral rectus muscle

Answer: turns eyeball laterally via lateral rectus muscle
Type: MA   Page Ref: 581
Topic: CRANIAL NERVES

23) Column 1: facial VII
Column 2: conveys impulses from the
taste buds on the anterior
two-thirds of the tongue

Answer: conveys impulses from the taste buds on the anterior two-thirds of the tongue
Type: MA   Page Ref: 581
Topic: CRANIAL NERVES

24) Column 1: vestibulocochlear VIII
Column 2: conveys impulses associated
with hearing and
equilibrium

Answer: conveys impulses associated with hearing and equilibrium
Type: MA   Page Ref: 581
Topic: CRANIAL NERVES

25) Column 1: glossopharyngeal IX
Column 2: conveys impulses from taste
buds on the posterior one-
third of the tongue

Answer: conveys impulses from taste buds on the posterior one-third of the tongue
Type: MA   Page Ref: 582
Topic: CRANIAL NERVES

26) Column 1: vagus X
Column 2: conveys impulses to
visceral, cardiac, and
skeletal muscles

Answer: conveys impulses to visceral, cardiac, and skeletal muscles
Type: MA   Page Ref: 582
Topic: CRANIAL NERVES

27) Column 1: accessory XI
Column 2: conveys motor impulses to
laryngeal and pharyngeal
muscles, and to the
sternocleidomastoid and
trapezius muscles

Answer: conveys motor impulses to laryngeal and pharyngeal muscles, and to the
sternocleidomastoid and trapezius muscles
Type: MA   Page Ref: 583
Topic: CRANIAL NERVES

28) Column 1: hypoglossal XII
Column 2: controls movement of the
tongue during speech,
chewing, and swallowing

Answer: controls movement of the tongue during speech, chewing, and swallowing
Type: MA   Page Ref: 583
Topic: CRANIAL NERVES

CHAPTER 19 General Senses and Sensory and Motor Pathways

MULTIPLE CHOICE. Choose the one alternative that best completes the statement or answers the question.

1) Which of the following pairs of terms is *incorrectly* matched?
   A) exteroceptors; at or near the surface of the body
   B) proprioceptors; in walls of blood vessels
   C) cutaneous receptors; in skin and mucous membranes
   D) interoceptors; in internal organs (viscerae)

   Answer: B
   Type: MC   Page Ref: 593
   Topic: SENSATION

2) Sensations may be translated at all of the following levels of the CNS. Perception, however, occurs when sensory impulses reach which region?
   A) spinal cord   B) brain stem   C) thalamus   D) cerebral cortex

   Answer: D
   Type: MC   Page Ref: 592
   Topic: SENSATION

3) The conversion of a stimulus into a nerve impulse is called:
   A) stimulation.   B) transduction.   C) conduction.   D) translation.

   Answer: B
   Type: MC   Page Ref: 592
   Topic: SENSATION

4) Sensory nerve impulses from the retina arrive in the visual cortex in the occipital lobes where they are interpreted as specific visual stimuli. The process occurring in the visual cortex is:
   A) stimulation.   B) transduction.   C) conduction.   D) translation.

   Answer: D
   Type: MC   Page Ref: 592
   Topic: SENSATION

5) The receptors for pain:
   A) are free nerve endings called nociceptors.
   B) may be stimulated by chemicals released by injured tissues.
   C) respond to any type of stimulus if it is of sufficient intensity.
   D) all of the above.

   Answer: D
   Type: MC   Page Ref: 595
   Topic: PAIN SENSATIONS

6) Which of the following cutaneous receptors are responsible for discriminative touch?
    1. type I cutaneous mechanoreceptors
    2. end organs of Ruffini
    3. type II cutaneous mechanoreceptors
    4. corpuscles of touch (Meissner's)
    A) 1, 2, 3, 4   B) 1, 3   C) 2, 4   D) 1, 4
    Answer: D
    Type: MC   Page Ref: 593, 595
    Topic: CUTANEOUS SENSATIONS

7) Which of the following are known to be free nerve endings?
    1. nociceptors
    2. thermoreceptors
    3. proprioceptors
    4. type I mechanoreceptors
    5. type II mechanoreceptors
    6. receptors for itch and tickle
    A) 2, 4, 6   B) 1, 2, 6   C) 3, 4, 5   D) 1, 2, 3
    Answer: B
    Type: MC   Page Ref: 593
    Topic: CUTANEOUS SENSATIONS

8) Which of the following is *not* a proprioceptor?
    A) hair root plexus   B) hair cells of the inner ear
    C) tendon organ   D) muscle spindle
    Answer: A
    Type: MC   Page Ref: 593, 597
    Topic: PROPRIOCEPTIVE SENSATIONS

9) Which of the following develop as specialized muscle fibers?
    A) tendon organs   B) hair cells of the inner ear
    C) muscle spindles   D) joint kinesthetic receptors
    Answer: C
    Type: MC   Page Ref: 597
    Topic: PROPRIOCEPTIVE SENSATIONS

10) From the following list of spinal cord pathways, choose the two that are together
    responsible for carrying all conscious sensory information in the spinal cord:
    1. spinothalamic
    2. fasciculus gracilis
    3. spinocerebellar
    4. posterior column–medial lemniscus
    5. fasciculus cuneatus
    A) 1, 4   B) 2, 5   C) 3, 4   D) 1, 3
    Answer: A
    Type: MC   Page Ref: 598, 599
    Topic: SENSORY PATHWAYS

11) Third-order neurons:
    A) have axons that end in the thalamus.
    B) carry nerve impulses from the spinal cord and brain stem to the thalamus.
    C) have axons that decussate in the spinal cord or brain stem.
    D) carry nerve impulses to the primary somatosensory area in the postcentral gyrus.

    Answer: D
    Type: MC   Page Ref: 598
    Topic: SENSORY PATHWAYS

12) Which of the following is a sensation that is *not* carried by the spinothalamic pathways?
    A) pain   B) itch   C) tickle   D) proprioception

    Answer: D
    Type: MC   Page Ref: 599
    Topic: SENSORY PATHWAYS

13) Pyramidal pathways:
    A) innervate skeletal muscles only.
    B) contain upper motor neurons, lower motor neurons, and, in most cases, association neurons.
    C) extend from the cerebral cortex, through the medulla, and then to their destination, either through cranial nerves or through spinal cord tracts and spinal nerves.
    D) all of the above.

    Answer: D
    Type: MC   Page Ref: 603
    Topic: MOTOR PATHWAYS

14) Corticobulbar tracts carry information that controls the voluntary movement of all of the following *except*:
    A) facial muscles.   B) eyes.
    C) skilled movements of the hands.   D) muscles of speech and chewing.

    Answer: C
    Type: MC   Page Ref: 603
    Topic: MOTOR PATHWAYS

15) The group of tracts called the lateral corticospinal tracts, anterior corticospinal tracts, and corticobulbar tracts:
    A) are the direct motor pathways.
    B) are the pyramidal motor pathways.
    C) are the indirect motor pathways.
    D) are the direct and indirect motor pathways.

    Answer: A
    Type: MC   Page Ref: 603
    Topic: MOTOR PATHWAYS

16) Upper motor neurons of the indirect pathways begin in the:
A) cerebral cortex.   B) brain stem.   C) thalamus.   D) spinal cord.

Answer: B
Type: MC   Page Ref: 604
Topic: MOTOR PATHWAYS

17) Final common pathways consist of:
A) upper motor neurons.   B) lower motor neurons.
C) upper and lower motor neurons.   D) association neurons.

Answer: B
Type: MC   Page Ref: 603
Topic: MOTOR PATHWAYS

18) Lower motor neurons receive both excitatory and inhibitory input from which of the following:
1. upper motor neurons
2. association neurons
3. direct pathways
4. indirect pathways
A) 1, 2, 3, 4   B) 1, 3, 4   C) 1 and 3 only   D) 1 and 2 only

Answer: A
Type: MC   Page Ref: 603
Topic: MOTOR PATHWAYS

TRUE/FALSE. Write 'T' if the statement is true and 'F' if the statement is false.

1) Selectivity of receptors refers to the fact that a receptor will only respond if a stimulus is of sufficient strength.

Answer: FALSE
Type: TF   Page Ref: 592
Topic: SENSATION

2) All receptors for general senses are of the simple type.

Answer: TRUE
Type: TF   Page Ref: 593
Topic: SENSATION

3) Type II cutaneous mechanoreceptors, found deep in the dermis, function as receptors for heavy continuous touch as well as for pressure.

Answer: TRUE
Type: TF   Page Ref: 595
Topic: CUTANEOUS SENSATIONS

4) Visceral pain occurs when nociceptors in internal organs, skeletal muscles, tendons, or joints are stimulated.

Answer: FALSE
Type: TF   Page Ref: 595
Topic: PAIN SENSATIONS

5) Stimulation of nociceptors in the skin gives rise to superficial somatic pain.

Answer: TRUE
Type: TF   Page Ref: 595
Topic: PAIN SENSATIONS

6) Proprioception is the sense that provides an awareness of body positions and movements of parts of the body.

Answer: FALSE
Type: TF   Page Ref: 596
Topic: PROPRIOCEPTIVE SENSATIONS

7) The fasciculus gracilis and fasciculus cuneatus are found in the anterior column of the spinal cord.

Answer: FALSE
Type: TF   Page Ref: 598
Topic: SENSORY PATHWAYS

8) Proprioceptive information may be conveyed to the cerebellum via axon collaterals of sensory neurons that travel in spinocerebellar tracts.

Answer: TRUE
Type: TF   Page Ref: 600
Topic: SENSORY PATHWAYS

9) First-order neurons transmit signals from somatic receptors along cranial and spinal nerves into the brain stem and spinal cord.

Answer: TRUE
Type: TF   Page Ref: 597
Topic: SENSORY PATHWAYS

10) Decussation of sensory pathways occurs before the pathway reaches the thalamus.

Answer: TRUE
Type: TF   Page Ref: 598
Topic: SENSORY PATHWAYS

11) The majority of commands from the motor cortex to skeletal muscles travel via the anterior corticospinal tracts in the spinal cord.

Answer: FALSE
Type: TF   Page Ref: 603
Topic: MOTOR PATHWAYS

12) The rubrospinal, tectospinal, reticulospinal, and corticospinal tracts are all indirect pathways.

Answer: FALSE
Type: TF   Page Ref: 605
Topic: MOTOR PATHWAYS

ESSAY.   Write your answer in the space provided or on a separate sheet of paper.

1)   A mosquito bites you on the wrist. Even before you look at the spot, your CNS has informed you of what has happened, and where. What four events have occurred between the time the insect began to bite and the moment when you became aware?

Answer: 1. Stimulation: activation of sensory neurons
   2. Transduction: creation of a nerve impulse in a first-order sensory neuron
   3. Conduction: nerve impulse travels to the CNS via the first-order neuron
   4. Translation: nerve impulse reaches the region in the CNS where it is transformed into a sensation
Type: ES   Page Ref: 592
Topic: SENSATION

2)   Outline the two classification systems for receptors. List the types of receptors according to each classification and give examples.

Answer: 1. By location: exteroreceptors, interoreceptors, proprioceptors
   2. By type of stimulus: mechanoreceptors, photoreceptors, chemoreceptors, nociceptors
   Examples are given on pp. 593.
Type: ES   Page Ref: 593
Topic: SENSATION

3)   What is referred pain? Using an example, explain how referred pain occurs.

Answer: Pain impulses from the viscera are often interpreted as coming from the skin either near to, or more removed from, the affected organ. The most common example is the pain of a heart attack. The area to which the pain is referred (medial aspect of left arm, skin over heart) is served by the T1–T4 region of the cord, which also serves the heart.
Type: ES   Page Ref: 595
Topic: PAIN SENSATIONS

4)   Name three spinal tracts that carry conscious sensory information. For each tract, describe the location and the types of sensations transmitted.

Answer: 1. Posterior column–medial lemniscus
   2. and 3. Anterior spinothalamic and lateral spinothalamic
Type: ES   Page Ref: 598, 599
Topic: SENSORY PATHWAYS

5)   Name, locate, and briefly describe the functions of three descending tracts that are referred to as the direct pathways.

Answer: The corticospinal (anterior and lateral) and corticobulbar tracts are described on pp. 603.
Type: ES   Page Ref: 603
Topic: MOTOR PATHWAYS

SHORT ANSWER. Write the word or phrase that best completes each statement or answers the question.

1) There are two main ways to classify receptors: by type of stimulus and by _____.

Answer: location
Type: SA   Page Ref: 598
Topic: SENSATION

2) The change in level of sensitivity of a receptor to a long–lasting stimulus is called _____.

Answer: adaptation
Type: SA   Page Ref: 593
Topic: SENSATION

3) The stimulation of nociceptors in skeletal muscles and joints gives rise to _____ pain.

Answer: deep somatic
Type: SA   Page Ref: 595
Topic: PAIN SENSATIONS

4) There are two destinations in the brain for proprioceptive information: the _____ and the _____.

Answer: cerebrum (parietal lobe, postcentral gyrus), cerebellum
Type: SA   Page Ref: 597
Topic: PROPRIOCEPTIVE SENSATIONS

5) _____ are proprioceptors located at the junction of a tendon and a muscle.

Answer: tendon organs
Type: SA   Page Ref: 597
Topic: PROPRIOCEPTIVE SENSATIONS

6) The cell bodies of _____-order sensory neurons are located in the thalamus.

Answer: second
Type: SA   Page Ref: 598
Topic: SENSORY PATHWAYS

MATCHING. Choose the item in Column 2 that best matches each item in Column 1.

Match the indirect pathways in Column 1 with the region of the brain stem in which they originate in Column 2.

1) Column 1: rubrospinal
Column 2: midbrain

Answer: midbrain
Type: MA   Page Ref: 604
Topic: MOTOR PATHWAYS

2) Column 1: tectospinal
Column 2: midbrain

Answer: midbrain
Type: MA   Page Ref: 604
Topic: MOTOR PATHWAYS

3) Column 1: vestibulospinal
Column 2: medulla

Answer: medulla
Type: MA   Page Ref: 604
Topic: MOTOR PATHWAYS

4) Column 1: lateral reticulospinal
Column 2: medulla

Answer: medulla
Type: MA   Page Ref: 604
Topic: MOTOR PATHWAYS

5) Column 1: medial reticulospinal
Column 2: pons

Answer: pons
Type: MA   Page Ref: 604
Topic: MOTOR PATHWAYS

Match the proprioceptors in Column 1 with their descriptions in Column 2.
6) Column 1: muscle spindle
Column 2: contains intrafusal fibers

Answer: contains intrafusal fibers
Type: MA   Page Ref: 597
Topic: PROPRIOCEPTIVE SENSATIONS

7) Column 1: muscle spindle
Column 2: innervated by small-
          diameter gamma motor
          neurons

Answer: innervated by small-diameter gamma motor neurons
Type: MA   Page Ref: 597
Topic: PROPRIOCEPTIVE SENSATIONS

8) Column 1: muscle spindle
Column 2: contains dendrites of type I
          or type II sensory fibers

Answer: contains dendrites of type I or type II sensory fibers
Type: MA   Page Ref: 597
Topic: PROPRIOCEPTIVE SENSATIONS

9) Column 1: tendon organ
Column 2: contains dendrites of type
          Ib sensory fibers

Answer: contains dendrites of type Ib sensory fibers
Type: MA   Page Ref: 597
Topic: PROPRIOCEPTIVE SENSATIONS

10) Column 1: tendon organ
Column 2: a thin capsule of connective
            tissue enclosing a few
            collagen fibers

Answer: a thin capsule of connective tissue enclosing a few collagen fibers
Type: MA   Page Ref: 597
Topic: PROPRIOCEPTIVE SENSATIONS

Match the tracts in Column 1 with their functions in Column 2.
11) Column 1: posterior column–medial
            lemniscus
Column 2: impulses for conscious
            proprioception and most
            tactile sensations

Answer: impulses for conscious proprioception and most tactile sensations
Type: MA   Page Ref: 602
Topic: SENSORY PATHWAYS

12) Column 1: lateral spinothalamic
Column 2: impulses for pain and
            temperature sensations

Answer: impulses for pain and temperature sensations
Type: MA   Page Ref: 602
Topic: SENSORY PATHWAYS

13) Column 1: anterior spinothalamic
Column 2: impulses for itch, tickle,
            crude touch, and pressure
            sensations

Answer: impulses for itch, tickle, crude touch, and pressure sensations
Type: MA   Page Ref: 602
Topic: SENSORY PATHWAYS

14) Column 1: anterior and posterior
            spinocerebellar
Column 2: impulses for subconscious
            proprioception

Answer: impulses for subconscious proprioception
Type: MA   Page Ref: 602
Topic: SENSORY PATHWAYS

15) Column 1: lateral corticospinal
Column 2: impulses for coordination of
            precise voluntary
            movements of the hands
            and feet

Answer: impulses for coordination of precise voluntary movements of the hands and feet
Type: MA   Page Ref: 605
Topic: MOTOR PATHWAYS

16) Column 1: anterior corticospinal
Column 2: impulses for coordination of
movements of the axial
skeleton

Answer: impulses for coordination of movements of the axial skeleton
Type: MA   Page Ref: 605
Topic: MOTOR PATHWAYS

17) Column 1: tectospinal
Column 2: impulses that move the
head and eyes in response
to visual stimuli

Answer: impulses that move the head and eyes in response to visual stimuli
Type: MA   Page Ref: 605
Topic: MOTOR PATHWAYS

18) Column 1: corticobulbar
Column 2: impulses that control
voluntary movement of the
eyes, tongue, neck, jaw, and
facial muscles via cranial
nerves III to VII and IX to
XII

Answer: impulses that control voluntary movement of the eyes, tongue, neck, jaw, and
facial muscles via cranial nerves III to VII and IX to XII
Type: MA   Page Ref: 605
Topic: MOTOR PATHWAYS

MULTIPLE CHOICE. Choose the one alternative that best completes the statement or answers the question.

1)   Mucus found on the surface of the olfactory epithelium:
A) is produced by glands beneath the olfactory epithelium.
B) makes it possible for odorant gases to react with the olfactory receptors.
C) is continuously renewed and therefore serves to cleanse the area of odorant gases, allowing new ones to act as stimuli.
D) all of the above.

Answer: D
Type: MC   Page Ref: 611
Topic: OLFACTORY SENSATIONS

2)   The olfactory epithelium occupies the superior part of the:
A) nasal cavity.   B) nasal septum.
C) nasal conchae.   D) all of the above.

Answer: D
Type: MC   Page Ref: 611
Topic: OLFACTORY SENSATIONS

3)   Which of the following is true for olfactory hairs?
A) They are cilia of olfactory neurons.
B) They are dendrites of olfactory neurons.
C) They are cilia of the supporting cells.
D) They produce mucus in the olfactory glands.

Answer: A
Type: MC   Page Ref: 611
Topic: OLFACTORY SENSATIONS

4)   Place the following in the correct order for the transmission of olfactory information that results in odor perception.
1. olfactory bulbs
2. olfactory tract
3. temporal cortex
4. olfactory nerves
5. frontal cortex
A) 4, 1, 2, 3, 5   B) 4, 2, 1, 3, 5   C) 2, 1, 4, 3, 5   D) 2, 4, 1, 3, 5

Answer: A
Type: MC   Page Ref: 611
Topic: OLFACTORY SENSATIONS

5)   Taste buds are located on the:
A) larynx.   B) pharynx.
C) tongue and soft palate.   D) all of the above.

Answer: D
Type: MC   Page Ref: 612
Topic: GUSTATORY SENSATIONS

6) The cranial nerves containing neurons of the gustatory
   pathway are:
   1. facial VII
   2. vagus X
   3. glossopharyngeal IX
   4. trigeminal V
   5. abducens VI

   A) 1, 3, 5   B) 2, 3, 4   C) 1, 2, 3   D) 1, 4, 5

   Answer: C
   Type: MC   Page Ref: 614
   Topic: GUSTATORY SENSATIONS

7) Which of the following is *not* an accessory structure of the eye?
   A) palpebrae   B) lacrimal apparatus
   C) intrinsic muscles   D) eyelashes

   Answer: C
   Type: MC   Page Ref: 614
   Topic: VISUAL SENSATIONS

8) A layer of dense connective tissue that covers all of the eyeball except the cornea is
   called _____.
   A) choroid   B) conjunctiva   C) sclera   D) uvea

   Answer: C
   Type: BI   Page Ref: 616
   Topic: VISUAL SENSATIONS

9) Which of the following types of cells is the first to receive a light ray as it travels
   through the retina?
   A) bipolar cell   B) ganglion cell
   C) pigment epithelium   D) photoreceptor cell

   Answer: B
   Type: MC   Page Ref: 618
   Topic: VISUAL SENSATIONS

10) The vitreous body:
    A) is produced by the ciliary processes.
    B) undergoes constant replacement.
    C) is the main contributor to intraocular pressure.
    D) none of the above.

    Answer: D
    Type: MC   Page Ref: 621
    Topic: VISUAL SENSATIONS

11) Which of the following is *not* a feature of the external ear?
    A) tympanic antrum   B) helix
    C) lobule   D) tympanic membrane

    Answer: A
    Type: MC   Page Ref: 624
    Topic: AUDITORY SENSATIONS AND EQUILIBRIUM

12) In what order do the following vibrate as sound is transmitted?
1. stapes
2. tympanic membrane
3. oval window
4. incus
5. malleus

A) 5, 1, 4, 2, 3   B) 3, 4, 5, 1, 2   C) 2, 1, 5, 4, 3   D) 2, 5, 4, 1, 3

Answer: D
Type: MC   Page Ref: 624
Topic: AUDITORY SENSATIONS AND EQUILIBRIUM

13) Which of the following is primarily responsible for sensations of hearing?
A) spiral organ   B) ampulla   C) saccule   D) utricle

Answer: A
Type: MC   Page Ref: 626
Topic: AUDITORY SENSATIONS AND EQUILIBRIUM

14) The vestibular apparatus is located in the:
A) middle ear.   B) anterior cavity of the eye.
C) nasal cavity.   D) inner ear.

Answer: D
Type: MC   Page Ref: 630
Topic: AUDITORY SENSATIONS AND EQUILIBRIUM

TRUE/FALSE.   Write 'T' if the statement is true and 'F' if the statement is false.

1) Olfactory receptors, which are neurons, are replaced regularly by the supporting cells of the olfactory epithelium.

Answer: FALSE
Type: TF   Page Ref: 611
Topic: OLFACTORY SENSATIONS

2) Olfactory receptor cells are unipolar neurons.

Answer: FALSE
Type: TF   Page Ref: 611
Topic: OLFACTORY SENSATIONS

3) Large papillae that always contain taste buds and that are found in a V-shaped row on the posterior region of the tongue are called fungiform papillae.

Answer: FALSE
Type: TF   Page Ref: 613
Topic: GUSTATORY SENSATIONS

4) The mucous membrane lining of the eyelids is the bulbar or ocular conjunctiva.

Answer: FALSE
Type: TF   Page Ref: 614
Topic: VISUAL SENSATIONS

5) The tip of the tongue reacts to all four primary taste stimuli, but is most sensitive to sweet and salty substances.

Answer: TRUE
Type: TF   Page Ref: 613
Topic: GUSTATORY SENSATIONS

6) In the retina, rods are more numerous than cones.

Answer: TRUE
Type: TF   Page Ref: 618
Topic: VISUAL SENSATIONS

7) The space within the eye between the lens and the cornea is the anterior cavity.

Answer: TRUE
Type: TF   Page Ref: 620
Topic: VISUAL SENSATIONS

8) The bony labyrinth contains perilymph; the membranous labyrinth contains endolymph.

Answer: TRUE
Type: TF   Page Ref: 633
Topic: AUDITORY SENSATIONS AND EQUILIBRIUM

9) Stereocilia are microvilluslike extensions of the membrane of the hair cells of the maculae in the utricle and saccule.

Answer: TRUE
Type: TF   Page Ref: 633
Topic: AUDITORY SENSATIONS AND EQUILIBRIUM

10) Sensory input from organs of dynamic and static equilibrium is transmitted to the pons and cerebellum regions of the brain.

Answer: TRUE
Type: TF   Page Ref: 569
Topic: AUDITORY SENSATIONS AND EQUILIBRIUM

ESSAY. Write your answer in the space provided or on a separate sheet of paper.

1) Olfactory receptors and gustatory receptors each contain receptor cells, supporting cells, and basal cells. Briefly describe each of these cells and state the function of each. Your answer should point out some major differences in each cell type between the two receptors.

Answer: 1. Olfactory epithelium: olfactory cells: bipolar neurons derived from division of basal cells (stem cells). Olfactory hairs on the dendrites: the site of olfactory transduction. Supporting cells: columnar epithelium.
2. Taste buds: All three cells here are epithelial cells. Supporting cells form the capsule; basal cells divide to produce supporting cells, which specialize to become gustatory receptor cells each with an apical gustatory hair that reacts with taste stimuli. Receptor cells synapse with the dendrites of the gustatory neurons at the base of the taste bud.

Type: ES   Page Ref: 611, 612
Topic: OLFACTORY SENSATIONS

2) Describe the various pathways for gustatory sensations from different regions of the oral cavity and pharynx. Then follow the common pathway to the gustatory cortex.

Answer: 1. Facial nerve VII serves the anterior two-thirds of the tongue.
2. Glossopharyngeal nerve IX serves the posterior one-third of the tongue.
3. Vagus nerve X serves the pharynx and epiglottis.
All impulses arrive in the medulla, then project to the thalamus and thence to the primary gustatory cortex of the parietal lobe.

Type: ES   Page Ref: 614
Topic: GUSTATORY SENSATIONS

3) Describe the wall of the eyeball, from outermost to innermost layer.

Answer: The following should be correctly used in the answer:
1. fibrous tunic, consisting of sclera and cornea
2. vascular tunic (uvea), consisting of choroid, ciliary body, ciliary processes, ciliary muscle, and iris
3. nervous tunic (retina), consisting of pigment epithelium and the neural portion of the retina

Type: ES   Page Ref: 616–618
Topic: VISUAL SENSATIONS

4) Summarize the events that occur as sound waves enter the external ear, then are transmitted through the middle and inner ears to produce sensations of hearing. Accompany your summary with labeled sketches that illustrate the anatomical features in your answer.

Answer: See Mechanism of Hearing, pp. 627.

Type: ES   Page Ref: 627, 628
Topic: AUDITORY SENSATIONS AND EQUILIBRIUM

SHORT ANSWER. Write the word or phrase that best completes each statement or answers the question.

1) The olfactory I nerves terminate in masses of gray matter called _____, which are located on the inferior surface of the brain, superior to the ethmoid bone.

Answer: olfactory bulbs
Type: SA   Page Ref: 611
Topic: OLFACTORY SENSATIONS

2) Olfactory epithelium consists of three main kinds of cells: _____, _____, and _____.

Answer: olfactory receptors, supporting cells, basal cells
Type: SA   Page Ref: 611
Topic: OLFACTORY SENSATIONS

3) The olfactory I nerves consist of approximately 40 bundles of _____ of olfactory neurons that extend through foramina in the cribriform plate of the _____ bone.

Answer: axons, ethmoid
Type: SA   Page Ref: 611
Topic: OLFACTORY SENSATIONS

4) The sense of taste is known as _____.

Answer: gustation
Type: SA   Page Ref: 612
Topic: GUSTATORY SENSATIONS

5) The rough appearance of the upper surface of the tongue is due to the presence of _____.

Answer: papillae
Type: SA   Page Ref: 613
Topic: GUSTATORY SENSATIONS

6) The three types of cells that make a taste bud are all _____ tissue.

Answer: epithelial
Type: SA   Page Ref: 612
Topic: GUSTATORY SENSATIONS

7) Lacrimal fluid contains water, mucus, salts, and a protective enzyme called _____.

Answer: lysozyme
Type: SA   Page Ref: 616
Topic: VISUAL SENSATIONS

8) In bright light, the _____ muscles of the iris contract, causing a/an _____ (increase/decrease) in the size of the pupil.

Answer: circular, decrease
Type: SA   Page Ref: 618
Topic: VISUAL SENSATIONS

9) The elevation located in the medial commissure of the eye and that contains sudoriferous and sebaceous glands is called the _____.

Answer: lacrimal caruncle
Type: SA   Page Ref: 614
Topic: VISUAL SENSATIONS

10) The yellow spot at the center of the posterior portion of the retina, called the _____, contains a depression called the _____ which contains only cone photoreceptors.

Answer: macula lutea, central fovea
Type: SA   Page Ref: 618
Topic: VISUAL SENSATIONS

11) The auditory tube connects the _____ to the _____.

Answer: middle ear, nasopharynx
Type: SA   Page Ref: 624
Topic: AUDITORY SENSATIONS AND EQUILIBRIUM

12) The complex series of canals known as the labyrinth make up the _____ ear.

Answer: inner
Type: SA   Page Ref: 624
Topic: AUDITORY SENSATIONS AND EQUILIBRIUM

13) In the wall of the utricle and saccule, the gelatinous layer of the macula that rests on the hair cells contains calcium carbonate crystals called _____.

Answer: otoliths
Type: SA   Page Ref: 633
Topic: AUDITORY SENSATIONS AND EQUILIBRIUM

14) Nerve impulses from maculae and cristae of the inner ear are carried by the _____ cranial nerve.

Answer: vestibulocochlear VIII
Type: SA   Page Ref: 633
Topic: AUDITORY SENSATIONS AND EQUILIBRIUM

15) The organs responsible for static and dynamic equilibrium are collectively known as the _____ apparatus.

Answer: vestibular
Type: SA   Page Ref: 633
Topic: AUDITORY SENSATIONS AND EQUILIBRIUM

MATCHING.   Choose the item in Column 2 that best matches each item in Column 1.

Match the papillae in Column 1 with their descriptions in Column 2.
1)   Column 1: circumvallate
     Column 2: largest type; all contain
               taste buds

     Answer: largest type; all contain taste buds
     Type: MA   Page Ref: 613
     Topic: GUSTATORY SENSATIONS

2)   Column 1: circumvallate
     Column 2: distributed in V-shaped
               row on the posterior
               portion of the tongue

     Answer: distributed in V-shaped row on the posterior portion of the tongue
     Type: MA   Page Ref: 613
     Topic: GUSTATORY SENSATIONS

3)   Column 1: fungiform
     Column 2: mushroom shaped; evenly
               distributed over the tongue

     Answer: mushroom shaped; evenly distributed over the tongue
     Type: MA   Page Ref: 613
     Topic: GUSTATORY SENSATIONS

4)   Column 1: filiform
     Column 2: rarely contain taste buds

     Answer: rarely contain taste buds
     Type: MA   Page Ref: 613
     Topic: GUSTATORY SENSATIONS

5)   Column 1: filiform
      Column 2: pointed threadlike papillae,
                distributed over the entire
                surface of the tongue

      Answer: pointed threadlike papillae, distributed over the entire surface of the tongue
      Type: MA   Page Ref: 613
      Topic: GUSTATORY SENSATIONS

Match the glands in Column 1 with their location in Column 2.
6)   Column 1: ceruminous glands
      Column 2: near the external opening
                of the external auditory
                canal

      Answer: near the external opening of the external auditory canal
      Type: MA   Page Ref: 623
      Topic: AUDITORY SENSATIONS AND EQUILIBRIUM

7)   Column 1: tarsal (Meibomian) glands
      Column 2: embedded in tarsal plate of
                eyelid

      Answer: embedded in tarsal plate of eyelid
      Type: MA   Page Ref: 614
      Topic: VISUAL SENSATIONS

8)   Column 1: lacrimal glands
      Column 2: superior and lateral to the
                eyeball

      Answer: superior and lateral to the eyeball
      Type: MA   Page Ref: 614
      Topic: VISUAL SENSATIONS

9)   Column 1: olfactory (Bowman's) glands
      Column 2: in connective tissue
                underneath the olfactory
                epithelium

      Answer: in connective tissue underneath the olfactory epithelium
      Type: MA   Page Ref: 611
      Topic: OLFACTORY SENSATIONS

CHAPTER 21   The Autonomic Nervous System

MULTIPLE CHOICE. Choose the one alternative that best completes the statement or answers the question.

1)   Which of the following statements is true for the autonomic nervous system?
   A) Afferent signals are usually consciously perceived.
   B) The motor neurons are always excitatory.
   C) Autonomic motor neurons secrete only one neurotransmitter.
   D) None of the above.

   Answer: D
   Type: MC   Page Ref: 640
   Topic: COMPARISON OF SNS AND ANS

2)   Which of the following statements about the ANS is false?
   A) Structurally, the ANS consists of two main components: a sensory (input) component and a motor (output) component.
   B) ANS motor (output) pathways contain two motor neurons in series.
   C) The motor (output) division of the ANS has three divisions: parasympathetic, sympathetic, and somatic.
   D) ANS stimulation results in excitation or inhibition of smooth muscle or cardiac muscle, or a change in secretory activity in glands.

   Answer: C
   Type: MC   Page Ref: 640
   Topic: COMPARISON OF SNS AND ANS

3)   The synapse between the motor neurons of a somatic nervous system pathway occurs in the:
   A) brain.   B) spinal cord.
   C) peripheral ganglia.   D) either A or B.

   Answer: D
   Type: MC   Page Ref: 640
   Topic: COMPARISON OF SNS AND ANS

4)   The synapse between the motor neurons of an autonomic motor pathway occurs in the:
   A) brain.   B) spinal cord.   C) PNS.   D) either A or B.

   Answer: C
   Type: MC   Page Ref: 640
   Topic: COMPARISON OF SNS AND ANS

5)   The somatic nervous system contains both sensory and motor neurons. The autonomic nervous system contains only autonomic motor neurons.
   A) Both statements are true.
   B) Both statements are false.
   C) The first statement is true; the second is false.
   D) The second statement is true; the first is false.

   Answer: C
   Type: MC   Page Ref: 640
   Topic: COMPARISON OF SNS AND ANS

6) A preganglionic fiber:
A) is a myelinated axon
B) is an unmyelinated axon
C) terminates in smooth muscle, cardiac muscle, or a gland
D) both A and C are correct

Answer: A
Type: MC   Page Ref: 641
Topic: AUTONOMIC MOTOR PATHWAYS

7) The anterior root of a thoracic spinal nerve contains which of the following?
1. somatic motor fibers
2. sympathetic preganglionic fibers
3. parasympathetic preganglionic fibers
A) 1, 2, 3   B) 1, 3   C) 1, 2   D) 2, 3

Answer: C
Type: MC   Page Ref: 645
Topic: SYMPATHETIC DIVISION

8) Place the following terms in the order in which a sympathetic motor impulse could travel:
1. gray ramus
2. anterior root of spinal nerve
3. axon collaterals
4. white ramus
5. nerve to a visceral effector
6. sympathetic trunk ganglion
A) 2, 4, 6, 3, 1, 5   B) 2, 1, 6, 3, 4, 5   C) 2, 3, 1, 6, 4, 5   D) 2, 3, 4, 6, 1, 5

Answer: A
Type: MC   Page Ref: 645
Topic: SYMPATHETIC DIVISION

9) White rami are associated only with thoracic spinal nerves and the first two or three lumbar spinal nerves. Gray rami are associated with all 31 pairs of spinal nerves.
A) Both statements are true.
B) Both statements are false.
C) The first statement is true; the second is false.
D) The second statement is true; the first is false.

Answer: A
Type: MC   Page Ref: 645
Topic: SYMPATHETIC DIVISION

10) Which of the following is *not* part of the sympathetic division?
A) gray rami   B) splanchnic nerves
C) solar plexus   D) submandibular ganglion

Answer: D
Type: MC   Page Ref: 647
Topic: AUTONOMIC MOTOR PATHWAYS

11) Which of the following is *not* a parasympathetic terminal ganglion?
A) pterygopalatine   B) otic   C) celiac   D) submandibular

Answer: C
Type: MC   Page Ref: 647
Topic: AUTONOMIC MOTOR PATHWAYS

12) Which of the following responses is a result of sympathetic stimulation?
A) decreased heart rate   B) pupillary constriction
C) increased digestive secretions   D) increased rate and depth of breathing

Answer: B
Type: MC   Page Ref: 650
Topic: ACTIVITIES OF THE ANS

13) Autonomic neurons that secrete the neurotransmitter acetylcholine:
A) are called adrenergic.
B) include all postganglionic neurons.
C) include all preganglionic neurons.
D) all sympathetic postganglionic neurons.

Answer: C
Type: MC   Page Ref: 648
Topic: ACTIVITIES OF THE ANS

14) Acetylcholine in the autonomic nervous system:
A) is found only in the parasympathetic division.
B) has a more prolonged effect than it does in the somatic nervous system.
C) has effects for a very brief time, compared to the effects of other autonomic neurotransmitters.
D) can have either excitatory or inhibitory effects.

Answer: C
Type: MC   Page Ref: 648
Topic: ACTIVITIES OF THE ANS

15) The neurotransmitter acetylcholine binds with which of the following types of receptors?
1. alpha
2. beta
3. nicotinic
4. muscarinic
A) 1, 2, 3, 4   B) 1, 2   C) 3, 4   D) 3 only

Answer: C
Type: MC   Page Ref: 649
Topic: ACTIVITIES OF THE ANS

16) Activation of nicotinic receptors of postganglionic neurons leads to:
A) excitation.
B) inhibition.
C) sometimes excitation, sometimes inhibition.
D) release of neurotransmitters.

Answer: A
Type: MC   Page Ref: 650
Topic: ACTIVITIES OF THE ANS

17) Alpha and beta receptors are found in the sympathetic division. They exist on postganglionic neurons and on visceral effectors.
A) Both statements are true.
B) Both statements are false.
C) The first statement is true; the second is false.
D) The second statement is true; the first is false.

Answer: C
Type: MC   Page Ref: 650
Topic: ACTIVITIES OF THE ANS

18) Which of the following is/are innervated only by the sympathetic division?
1. arrector pili muscles
2. blood vessels in the kidney
3. sweat glands
4. adipose tissue
5. skeletal muscle arterioles
6. brain arterioles
7. adrenal medulla

A) 1, 2, 3, 4, 5, 6, 7   B) 2, 3, 5, 6, 7
C) 1, 3, 4, 7   D) 2, 4, 5, 6

Answer: A
Type: MC   Page Ref: 651, 652
Topic: ACTIVITIES OF THE ANS

19) Arrange the following components of an autonomic reflex arc in the proper order:
1. preganglionic neuron
2. receptor
3. association neurons
4. postganglionic neuron
5. visceral effector
6. autonomic ganglion
7. sensory neuron

A) 7, 2, 1, 3, 4, 6, 5   B) 1, 2, 7, 3, 6, 5, 4
C) 2, 1, 3, 7, 4, 6, 5   D) 2, 7, 3, 1, 6, 4, 5

Answer: D
Type: MC   Page Ref: 653
Topic: ACTIVITIES OF THE ANS

TRUE/FALSE. Write 'T' if the statement is true and 'F' if the statement is false.

1) Dual innervation means that an organ is innervated by both the SNS and the ANS.

Answer: FALSE
Type: TF   Page Ref: 640
Topic: COMPARISON OF SNS AND ANS

2) The sympathetic division of the ANS has craniosacral outflow; the parasympathetic division has thoracolumbar outflow.

Answer: FALSE
Type: TF   Page Ref: 641
Topic: AUTONOMIC MOTOR PATHWAYS

3) Sympathetic preganglionic fibers are, in general, longer than parasympathetic preganglionic fibers.

Answer: FALSE
Type: TF   Page Ref: 645
Topic: AUTONOMIC MOTOR PATHWAYS

4) The cell bodies of ANS postganglionic fibers are located in autonomic ganglia.

Answer: TRUE
Type: TF   Page Ref: 645
Topic: AUTONOMIC MOTOR PATHWAYS

5) The celiac ganglion and the mesenteric ganglion are examples of prevertebral ganglia.

Answer: TRUE
Type: TF   Page Ref: 645
Topic: SYMPATHETIC DIVISION

6) Parasympathetic preganglionic neuron cell bodies are located in the brain stem and in lateral gray horns of the second to fourth sacral segments.

Answer: TRUE
Type: TF   Page Ref: 647
Topic: PARASYMPATHETIC DIVISION

7) Parasympathetic postganglionic fibers are very short because the terminal ganglia are very close to the structures they innervate.

Answer: TRUE
Type: TF   Page Ref: 647
Topic: PARASYMPATHETIC DIVISION

8) The lacrimal glands are not dually innervated; they receive only parasympathetic innervation.

Answer: TRUE
Type: TF   Page Ref: 651
Topic: PARASYMPATHETIC DIVISION

9) Nicotinic receptors and muscarinic receptors both interact with acetylcholine.

Answer: TRUE
Type: TF   Page Ref: 649
Topic: ACTIVITIES OF THE ANS

10) Sympathetic postganglionic fibers diverge more than parasympathetic postganglionic fibers do.

Answer: TRUE
Type: TF   Page Ref: 647
Topic: ACTIVITIES OF THE ANS

11) Adrenergic alpha receptors are usually excitatory.

Answer: TRUE
Type: TF   Page Ref: 650
Topic: ACTIVITIES OF THE ANS

12) Nerve impulses from the hypothalamus and reticular formation influence the activity of sympathetic nuclei in the brain stem and spinal cord.

Answer: TRUE
Type: TF   Page Ref: 653
Topic: ACTIVITIES OF THE ANS

ESSAY. Write your answer in the space provided or on a separate sheet of paper.

1) List the four cranial nerves that carry parasympathetic motor impulses. Three pairs of these nerves lead to four pairs of ganglia. Name the ganglia and state which cranial nerves connect to them.

Answer: Oculomotor (III), facial (VII), glossopharyngeal (IX), and vagus (X) carry parasympathetic impulses. The first three listed lead to ganglia as shown in Fig. 21.2.
Type: ES   Page Ref: 643, 647
Topic: PARASYMPATHETIC DIVISION

2) Which division of the ANS is responsible for preparing the body for an emergency situation? Describe the responses of the body that collectively are referred to as the fight-or-flight response.

Answer: The sympathetic division produces the fight-or-flight response. See list of items pp. 650.
Type: ES   Page Ref: 650-651
Topic: SYMPATHETIC DIVISION

3)   There are two types of receptors for both acetylcholine and norepinephrine. Give the names of the receptors, the general location of each, and the expected effect of stimulation of each receptor type.

Answer: Receptors for ACh are cholinergic receptors. The two types:
1. nicotinic receptors, located on the dendrites and cell bodies of sympathetic and parasympathetic postganglionic neurons, lead to excitation of the postsynaptic cell when activated.
2. muscarinic receptors, located on all effectors served by parasympathetic postganglionic neurons, cause excitation of some cells (as in the constriction of the iris), or inhibition of others (as in decreased heart rate and force of contraction) when stimulated.
Receptors for NE are adrenergic receptors. The two types are alpha receptors and beta receptors. Both are found on visceral effectors innervated by most sympathetic postganglionic neurons. Alpha receptors tend to be excitatory (as in dilation of the pupil). Beta receptors are excitatory in some locations (as in increased heart rate and force of contraction) and inhibitory in others (as in airway dilation).
Type: ES   Page Ref: 649-650
Topic: ACTIVITIES OF THE ANS

SHORT ANSWER. Write the word or phrase that best completes each statement or answers the question.

1)   An organ that is served by both divisions of the ANS is said to have _____.
Answer: dual innervation
Type: SA   Page Ref: 640
Topic: COMPARISON OF SNS AND ANS

2)   The output component of the ANS has two divisions: _____ and _____.
Answer: sympathetic, parasympathetic
Type: SA   Page Ref: 640
Topic: COMPARISON OF SNS AND ANS

3)   In almost all cases, synapses between preganglionic and postganglionic neurons of the sympathetic division occur either in _____ ganglia or in _____ ganglia.
Answer: prevertebral, paravertebral
Type: SA   Page Ref: 642
Topic: SYMPATHETIC DIVISION

4)   The _____ portion of the sympathetic trunk innervates the heart, sweat glands of the head, the smooth muscle of the eye, and blood vessels of the facial region.
Answer: cervical
Type: SA   Page Ref: 645
Topic: SYMPATHETIC DIVISION

5)   The facial VII cranial nerves carry parasympathetic preganglionic fibers to two pairs of terminal ganglia: the _____ and the _____ ganglia.
Answer: ciliary, pterygopalatine
Type: SA   Page Ref: 647
Topic: PARASYMPATHETIC DIVISION

6) The lateral gray horns of the thoracic region contain cell bodies of _____ (sympathetic or parasympathetic) preganglionic neurons.

Answer: sympathetic
Type: SA   Page Ref: 641
Topic: SYMPATHETIC DIVISION

7) There are _____ ganglia in each chain of the sympathetic trunk: _____ cervical, _____ thoracic, _____ lumbar, and _____ sacral.

Answer: 22, 3, 11, 4, 4
Type: SA   Page Ref: 642
Topic: SYMPATHETIC DIVISION

8) Splanchnic nerves carry sympathetic fibers from a/an _____ ganglion to a/an _____ ganglion.

Answer: paravertebral, prevertebral
Type: SA   Page Ref: 647
Topic: SYMPATHETIC DIVISION

9) The cranial nerve that carries most of the parasympathetic outflow is the _____.

Answer: vagus X
Type: SA   Page Ref: 647
Topic: PARASYMPATHETIC DIVISION

10) Most blood vessels are not dually innervated; they receive only _____ innervation.

Answer: sympathetic
Type: SA   Page Ref: 648
Topic: ACTIVITIES OF THE ANS

11) Autonomic neurons that release the neurotransmitter norepinephrine are called _____ neurons.

Answer: adrenergic
Type: SA   Page Ref: 648
Topic: ACTIVITIES OF THE ANS

12) The _____ division of the ANS is mostly responsible for regulating restorative, energy-conserving body activities.

Answer: parasympathetic
Type: SA   Page Ref: 650
Topic: ACTIVITIES OF THE ANS

13) In the CNS, the region called the _____is the major control area for the ANS.

Answer: hypothalamus
Type: SA   Page Ref: 653
Topic: ACTIVITIES OF THE ANS

MATCHING. Choose the item in Column 2 that best matches each item in Column 1.

Match the features of the autonomic nervous system in Column 1 with their descriptions in Column 2.

1)   Column 1: white ramus
      Column 2: contains sympathetic
              preganglionic axons

      Answer: contains sympathetic preganglionic axons
      Type: MA   Page Ref: 645
      Topic: AUTONOMIC MOTOR PATHWAYS

2)   Column 1: middle cervical ganglion
      Column 2: sympathetic; sends impulses
              along postganglionic axons
              to heart

      Answer: sympathetic; sends impulses along postganglionic axons to heart
      Type: MA   Page Ref: 645
      Topic: AUTONOMIC MOTOR PATHWAYS

3)   Column 1: celiac ganglion
      Column 2: receives sympathetic
              preganglionic fibers via
              thoracic splanchnic nerves

      Answer: receives sympathetic preganglionic fibers via thoracic splanchnic nerves
      Type: MA   Page Ref: 647
      Topic: AUTONOMIC MOTOR PATHWAYS

4)   Column 1: thoracic and lumbar lateral
              gray horns
      Column 2: contain sympathetic
              preganglionic cell bodies

      Answer: contain sympathetic preganglionic cell bodies
      Type: MA   Page Ref: 645
      Topic: AUTONOMIC MOTOR PATHWAYS

5)   Column 1: nuclei in brain stem
      Column 2: contain parasympathetic
              preganglionic cell bodies

      Answer: contain parasympathetic preganglionic cell bodies
      Type: MA   Page Ref: 647
      Topic: AUTONOMIC MOTOR PATHWAYS

6)   Column 1: vagus X
     Column 2: carries parasympathetic
             impulses to terminal
             ganglia in most organs of
             the ventral cavity
     Answer: carries parasympathetic impulses to terminal ganglia in most organs of the
             ventral cavity
     Type: MA   Page Ref: 647
     Topic: AUTONOMIC MOTOR PATHWAYS

7)   Column 1: otic ganglion
     Column 2: sends impulses via
             parasympathetic
             postganglionic fibers to the
             parotid gland
     Answer: sends impulses via parasympathetic postganglionic fibers to the parotid gland
     Type: MA   Page Ref: 617
     Topic: AUTONOMIC MOTOR PATHWAYS

Match the ANS preganglionic and postganglionic neurons in Column 1 with their descriptions in
Column 2.
8)   Column 1: sympathetic preganglionic
             neuron
     Column 2: cell bodies are in the
             lateral gray horns of all
             thoracic and first two or
             three lumbar spinal
             segments
     Answer: cell bodies are in the lateral gray horns of all thoracic and first two or three
             lumbar spinal segments
     Type: MA   Page Ref: 645
     Topic: AUTONOMIC MOTOR PATHWAYS

9)   Column 1: sympathetic preganglionic
             neuron
     Column 2: axons tend to be short
     Answer: axons tend to be short
     Type: MA   Page Ref: 645
     Topic: AUTONOMIC MOTOR PATHWAYS

10) Column 1: sympathetic postganglionic
             neuron
     Column 2: cell bodies are in vertebral
             chain ganglia or
             prevertebral ganglia
     Answer: cell bodies are in vertebral chain ganglia or prevertebral ganglia
     Type: MA   Page Ref: 647
     Topic: AUTONOMIC MOTOR PATHWAYS

11)   Column 1: parasympathetic
              preganglionic neuron
      Column 2: cell bodies are in the brain
              stem nuclei of the
              oculomotor III, facial VII,
              glossopharyngeal IX, and
              vagus X nerves

      Answer: cell bodies are in the brain stem nuclei of the oculomotor III, facial VII,
              glossopharyngeal IX, and vagus X nerves
      Type: MA   Page Ref: 647
      Topic: AUTONOMIC MOTOR PATHWAYS

12)   Column 1: parasympathetic
              preganglionic neuron
      Column 2: axons tend to be long

      Answer: axons tend to be long
      Type: MA   Page Ref: 647
      Topic: AUTONOMIC MOTOR PATHWAYS

13)   Column 1: parasympathetic
              postganglionic neuron
      Column 2: cell bodies are in terminal
              ganglia

      Answer: cell bodies are in terminal ganglia
      Type: MA   Page Ref: 647
      Topic: AUTONOMIC MOTOR PATHWAYS

Match the ANS neurons in Column 1 with their descriptions in Column 2.
14)   Column 1: cholinergic neurons
      Column 2: sympathetic preganglionic
              neurons

      Answer: sympathetic preganglionic neurons
      Type: MA   Page Ref: 648
      Topic: ACTIVITIES OF THE ANS

15)   Column 1: cholinergic neurons
      Column 2: parasympathetic
              preganglionic neurons

      Answer: parasympathetic preganglionic neurons
      Type: MA   Page Ref: 648
      Topic: ACTIVITIES OF THE ANS

16)   Column 1: cholinergic neurons
      Column 2: parasympathetic
              postganglionic neurons

      Answer: parasympathetic postganglionic neurons
      Type: MA   Page Ref: 648
      Topic: ACTIVITIES OF THE ANS

17) Column 1: adrenergic neurons
Column 2: release norepinephrine or
epinephrine

Answer: release norepinephrine or epinephrine
Type: MA   Page Ref: 648
Topic: ACTIVITIES OF THE ANS

18) Column 1: adrenergic neurons
Column 2: most sympathetic
postganglionic neurons

Answer: most sympathetic postganglionic neurons
Type: MA   Page Ref: 648
Topic: ACTIVITIES OF THE ANS

CHAPTER 22   The Endocrine System

MULTIPLE CHOICE. Choose the one alternative that best completes the statement or answers the question.

1)   Which of the following is considered exclusively an endocrine gland?
      A) hypothalamus   B) pituitary   C) adrenal   D) pancreas

      Answer: C
      Type: MC   Page Ref: 658
      Topic: ENDOCRINE GLANDS

2)   Which of the following is a space or fluid in which you would *not* expect to find a hormone?
      A) blood plasma   B) interstitial fluid   C) duct   D) capillaries

      Answer: C
      Type: MC   Page Ref: 658
      Topic: ENDOCRINE GLANDS

3)   The ability of a hormone to affect the activities of a particular cell depends on that cell having specific receptors with which the hormone molecules can bind. The number of receptors present in a target cell is a constant, invariable number.
      A) Both statements are true.
      B) Both statements are false.
      C) The first statement is true; the second is false.
      D) The second statement is true; the first is false.

      Answer: C
      Type: MC   Page Ref: 659
      Topic: HORMONES

4)   Which of the following hormones has the ovaries and testes as its target tissue?
      A) follicle–stimulating hormone   B) relaxin
      C) inhibin   D) progesterone

      Answer: A
      Type: MC   Page Ref: 661
      Topic: PITUITARY GLAND

5)   The _____ is the main link between the nervous system and the endocrine system, due to its control over secretory activities of the _____ gland.
      A) thalamus, pituitary   B) hypothalamus, thyroid
      C) brain stem, thyroid   D) hypothalamus, pituitary

      Answer: D
      Type: BI   Page Ref: 660
      Topic: PITUITARY GLAND

6)   The hypophyseal portal veins:
A) supply blood to the primary plexus capillaries in the base of the hypothalamus.
B) carry hypothalamic hormones to the posterior pituitary.
C) receive anterior pituitary hormones as they are secreted.
D) receive posterior pituitary hormones via the plexus of the infundibular process.

Answer: D
Type: MC   Page Ref: 660
Topic: PITUITARY GLAND

7)   If the anterior pituitary ceased to function, which of the following would *not* be directly affected?
A) mammary glands   B) kidneys
C) thyroid gland   D) adrenal cortex

Answer: B
Type: MC   Page Ref: 660, 661
Topic: PITUITARY GLAND

8)   The two lobes of the thyroid gland are joined by a mass of tissue called the _____.
A) isthmus   B) parathyroid   C) infundibulum   D) follicle

Answer: A
Type: BI   Page Ref: 664
Topic: THYROID GLAND

9)   Thyroxine is a hormone that:
A) is produced by parafollicular, or C, cells.
B) is secreted along with thyroglobulin into the follicles of the thyroid gland.
C) is also known as T_3 (triiodothyronine).
D) causes myxedema when produced in excess.

Answer: B
Type: MC   Page Ref: 664
Topic: THYROID GLAND

10)   Which of the adrenal gland hormones is/are normally secreted in minute quantities?
1. aldosterone
2. adrenocortical androgens
3. cortisol
A) 1, 2   B) 1, 3   C) 2 only   D) 3 only

Answer: C
Type: MC   Page Ref: 668
Topic: ADRENAL GLANDS

11) Epinephrine and norepinephrine from the adrenal medulla are sympathomimetic hormones. Sympathomimetic means the hormones are released in response to sympathetic stimulation.
   A) Both statements are true.
   B) Both statements are false.
   C) The first statement is true; the second is false.
   D) The second statement is true; the first is false.

   Answer: C
   Type: MC   Page Ref: 669
   Topic: ADRENAL GLANDS

12) Which of the following pairs of terms is incorrectly matched?
   A) insulin; secreted by beta cells
   B) glucagon; secreted by alpha cells
   C) somatostatin; secreted by delta cells
   D) pancreatic polypeptide; inhibits secretion of insulin and glucagon

   Answer: D
   Type: MC   Page Ref: 670
   Topic: PANCREAS

13) Estrogen is produced by cells in the:
   A) ovaries.   B) placenta.   C) adrenal cortex.   D) both A and B.

   Answer: D
   Type: MC   Page Ref: 672, 673
   Topic: OVARIES AND TESTES

14) The hormone inhibin is produced by:
   A) ovaries.   B) testes.   C) pancreas.   D) both A and B.

   Answer: D
   Type: MC   Page Ref: 672
   Topic: OVARIES AND TESTES

15) Which of the following is *not* part of the pancreatic islets (islets of Langerhans)?
   A) acini   B) alpha cells   C) beta cells   D) F-cells

   Answer: A
   Type: MC   Page Ref: 670
   Topic: PANCREAS

16) The hormone that promotes development and maintenance of male secondary sex characteristics is:
   A) follicle–stimulating hormone.   B) testosterone.
   C) luteinizing hormone.   D) human growth hormone.

   Answer: B
   Type: MC   Page Ref: 672
   Topic: OVARIES AND TESTES

17) The endocrine gland that produces hormones that mediate the proliferation and maturation of T cells of the immune system is the _____.
   A) pancreas   B) pineal   C) thymus   D) thyroid

   Answer: C
   Type: MC   Page Ref: 673
   Topic: THYMUS GLAND

18) The following gland(s) arise(s) from ectoderm:
   A) thyroid.   B) pancreas.
   C) thymus.   D) none of the above.

   Answer: D
   Type: MC   Page Ref: 674, 675
   Topic: DEVELOPMENTAL ANATOMY

TRUE/FALSE. Write 'T' if the statement is true and 'F' if the statement is false.

1) The nervous system responds to change very quickly, but an individual response is very brief; whereas the endocrine system's response takes longer to take effect but it lasts for a longer period of time.

   Answer: TRUE
   Type: TF   Page Ref: 658
   Topic: ENDOCRINE GLANDS

2) The anterior and posterior regions of the pituitary gland both develop from ectoderm.

   Answer: TRUE
   Type: TF   Page Ref: 660, 674
   Topic: PITUITARY GLAND

3) Oxyphil cells make up most of the volume of parathyroid glands and are primarily responsible for producing parathormone (PTH).

   Answer: FALSE
   Type: TF   Page Ref: 666
   Topic: PARATHYROID GLANDS

4) Parathormone is produced by the parafollicular cells of the thyroid gland.

   Answer: FALSE
   Type: TF   Page Ref: 666
   Topic: THYROID GLAND

5) Cortisol, the main glucocorticoid from the zona fasciculata, depresses the immune response and inflammation.

   Answer: TRUE
   Type: TF   Page Ref: 670
   Topic: ADRENAL GLANDS

6) Hormones of the adrenal medulla help prolong the fight-or-flight response.

   Answer: TRUE
   Type: TF   Page Ref: 669
   Topic: ADRENAL GLANDS

7) Endocrine portions of the pancreas are called acini.

Answer: FALSE
Type: TF   Page Ref: 670
Topic: PANCREAS

8) Thymosin and other thymic hormones influence the maturation of B cells in the thymus gland.

Answer: FALSE
Type: TF   Page Ref: 673
Topic: THYMUS GLAND

ESSAY. Write your answer in the space provided or on a separate sheet of paper.

1) List the five different types of secretory cells in the anterior pituitary together with the names of the hormones secreted by each.

Answer: 1. somatotrophs–hGH
      2. lactotrophs–PRL
      3. corticotrophs–ACTH and MSH
      4. thyrotrophs–TSH
      5. gonadotrophs–FSH and LH
Type: ES   Page Ref: 660, 661
Topic: PITUITARY GLAND

2) What is a tropic hormone? Give two examples secreted by the anterior pituitary.

Answer: A tropic hormone is one which has, as its target tissue, another endocrine gland. For example, FSH and LH from the anterior pituitary have the ovaries and testes as target tissues.
Type: ES   Page Ref: 662
Topic: PITUITARY GLAND

3) Name the cells that are responsible for producing the hormones of the posterior pituitary. Name the hormones together with their target tissues.

Answer: Posterior pituitary hormones are produced by the neurosecretory cells in the paraventricular and supraoptic nuclei of the hypothalamus. The axons of these cells form the supraopticohypophyseal tract, which carries the hormones into the posterior pituitary where they are stored in vesicles in the axon terminals. The two hormones and their target tissues are: OT (uterus and mammary glands), and ADH (kidneys).
Type: ES   Page Ref: 662, 664
Topic: PITUITARY GLAND

4) Name the secretory cell(s) of the thyroid gland, describe the location of the cell(s), list the hormone(s) produced, and state the target cells of the hormone(s).

Answer: Follicular cells in the walls of the follicles produce thyroxine (T_4) and triiodothyronine (T_3), which affect metabolism in almost all body cells. Parafollicular cells in the outer regions of the follicle walls, or between follicles, produce calcitonin which helps control blood calcium levels by increasing the utilization of calcium by osteoblasts.
Type: ES   Page Ref: 664
Topic: THYROID GLAND

SHORT ANSWER. Write the word or phrase that best completes each statement or answers the question.

1) The nervous and endocrine systems are sometimes referred to as one large system called the _____ system.
   Answer: neuroendocrine
   Type: SA   Page Ref: 658
   Topic: ENDOCRINE GLANDS

2) A hormone affects the activities of a few specific types of cells. Such cells are referred to as _____ cells.
   Answer: target
   Type: SA   Page Ref: 659
   Topic: HORMONES

3) The pituitary gland is attached to the hypothalamus by a stalk called the _____.
   Answer: infundibulum
   Type: SA   Page Ref: 660
   Topic: PITUITARY GLAND

4) The pituitary gland consists of two anatomically and functionally separate portions; the larger _____ and the smaller _____.
   Answer: anterior pituitary, posterior pituitary
   Type: SA   Page Ref: 660
   Topic: PITUITARY GLAND

5) Hypothalamic hormones travel via the primary plexus, the hypophyseal portal veins, and the secondary plexus, where they diffuse into the _____ pituitary.
   Answer: anterior
   Type: SA   Page Ref: 660
   Topic: PITUITARY GLAND

6) Parafollicular cells of the thyroid gland produce the hormone _____, which helps to regulate the level of _____ in the blood.
   Answer: calcitonin, calcium
   Type: SA   Page Ref: 664
   Topic: THYROID GLAND

7) Thyroid hormones result from enzyme–controlled reactions that combine the amino acid _____ with _____.
   Answer: tyrosine, iodine
   Type: SA   Page Ref: 664
   Topic: THYROID GLAND

8) Parathormone is produced by _____ cells of the _____ gland.
   Answer: principal or chief, parathyroid
   Type: SA   Page Ref: 666
   Topic: PARATHYROID GLANDS

9) The adrenal gland functions as two separate endocrine glands: the _____ and the _____.

Answer: adrenal cortex, adrenal medulla
Type: SA   Page Ref: 666
Topic: ADRENAL GLANDS

10) Hormones that regulate sodium and potassium levels in the blood are secreted by the _____ region of the adrenal cortex.

Answer: zona glomerulosa
Type: SA   Page Ref: 668
Topic: ADRENAL GLANDS

11) _____ cells of the adrenal gland have the same embryonic origin as all sympathetic postganglionic neurons.

Answer: chromaffin
Type: SA   Page Ref: 668
Topic: ADRENAL GLANDS

12) The pancreatic islets contain four types of hormone-producing cells: _____ cells, _____ cells, _____ cells, and _____ cells.

Answer: alpha, beta, delta, F–
Type: SA   Page Ref: 670
Topic: PANCREAS

13) The two hormones produced by the testes are _____ and _____.

Answer: testosterone, inhibin
Type: SA   Page Ref: 672
Topic: OVARIES AND TESTES

14) The _____ gland is located near the roof of the third ventricle of the brain.

Answer: pineal or epiphysis cerebri
Type: SA   Page Ref: 672
Topic: PINEAL GLAND

15) The pineal gland hormone that helps regulate the body's internal clock is _____.

Answer: melatonin
Type: SA   Page Ref: 672
Topic: PINEAL GLAND

MATCHING. Choose the item in Column 2 that best matches each item in Column 1.

Match the pituitary hormones in Column 1 with their target tissues in Column 2.

1) Column 1: PRL
Column 2: mammary glands

Answer: mammary glands
Type: MA   Page Ref: 664
Topic: PITUITARY GLAND

2)   Column 1: LH
        Column 2: ovaries, testes

        Answer: ovaries, testes
        Type: MA   Page Ref: 664
        Topic: PITUITARY GLAND

3)   Column 1: FSH
        Column 2: ovaries, testes

        Answer: ovaries, testes
        Type: MA   Page Ref: 664
        Topic: PITUITARY GLAND

4)   Column 1: TSH
        Column 2: thyroid gland

        Answer: thyroid gland
        Type: MA   Page Ref: 664
        Topic: PITUITARY GLAND

5)   Column 1: ACTH
        Column 2: adrenal cortex
        Foil: adrenal medulla

        Answer: adrenal cortex
        Type: MA   Page Ref: 664
        Topic: PITUITARY GLAND

6)   Column 1: hGH
        Column 2: bones and skeletal muscles

        Answer: bones and skeletal muscles
        Type: MA   Page Ref: 664
        Topic: PITUITARY GLAND

7)   Column 1: MSH
        Column 2: epidermis

        Answer: epidermis
        Type: MA   Page Ref: 664
        Topic: PITUITARY GLAND

Match the names of the hormone-secreting cells in Column 1 with their secretory products in Column 2.

8)   Column 1: follicular cells
        Column 2: thyroxine

        Answer: thyroxine
        Type: MA   Page Ref: 664
        Topic: THYROID GLAND

9)   Column 1: somatotrophs
        Column 2: human growth hormone

        Answer: human growth hormone
        Type: MA   Page Ref: 660
        Topic: PITUITARY GLAND

10) Column 1: principal cells
Column 2: parathormone

Answer: parathormone
Type: MA   Page Ref: 666
Topic: PARATHYROID GLANDS

11) Column 1: lactotrophs
Column 2: prolactin

Answer: prolactin
Type: MA   Page Ref: 661
Topic: PITUITARY GLAND

12) Column 1: parafollicular cells
Column 2: calcitonin

Answer: calcitonin
Type: MA   Page Ref: 664
Topic: THYROID GLAND

13) Column 1: chromaffin cells
Column 2: epinephrine

Answer: epinephrine
Type: MA   Page Ref: 668
Topic: ADRENAL GLANDS

14) Column 1: thyrotrophs
Column 2: TSH

Answer: TSH
Type: MA   Page Ref: 660
Topic: PITUITARY GLAND

15) Column 1: gonadotrophs
Column 2: FSH and LH

Answer: FSH and LH
Type: MA   Page Ref: 661
Topic: PITUITARY GLAND

16) Column 1: corticotrophs
Column 2: ACTH
Foil: cortisol

Answer: ACTH
Type: MA   Page Ref: 661
Topic: PITUITARY GLAND

17) Column 1: alpha cells
Column 2: glucagon

Answer: glucagon
Type: MA   Page Ref: 670
Topic: PANCREAS

18)   Column 1: beta cells
       Column 2: insulin

       Answer: insulin
       Type: MA   Page Ref: 670
       Topic: PANCREAS

19)   Column 1: delta cells
       Column 2: somatostatin

       Answer: somatostatin
       Type: MA   Page Ref: 670
       Topic: PANCREAS

20)   Column 1: F-cells
       Column 2: pancreatic polypeptide

       Answer: pancreatic polypeptide
       Type: MA   Page Ref: 670
       Topic: PANCREAS

21)   Column 1: pinealocytes
       Column 2: melatonin

       Answer: melatonin
       Type: MA   Page Ref: 672
       Topic: PINEAL GLAND

Match the regions of the adrenal cortex in Column 1 with their descriptions and hormones in Column 2.

22)   Column 1: zona reticularis
       Column 2: inner zone

       Answer: inner zone
       Type: MA   Page Ref: 668
       Topic: ADRENAL GLANDS

23)   Column 1: zona reticularis
       Column 2: secretes androgens

       Answer: secretes androgens
       Type: MA   Page Ref: 668
       Topic: ADRENAL GLANDS

24)   Column 1: zona glomerulosa
       Column 2: outer zone

       Answer: outer zone
       Type: MA   Page Ref: 666
       Topic: ADRENAL GLANDS

25)   Column 1: zona glomerulosa
       Column 2: secretes mineralocorticoids

       Answer: secretes mineralocorticoids
       Type: MA   Page Ref: 667
       Topic: ADRENAL GLANDS

26) Column 1: zona fasciculata
    Column 2: middle zone; also widest
             zone

    Answer: middle zone; also widest zone
    Type: MA   Page Ref: 667
    Topic: ADRENAL GLANDS

27) Column 1: zona fasciculata
    Column 2: secretes glucocorticoids

    Answer: secretes glucocorticoids
    Type: MA   Page Ref: 667
    Topic: ADRENAL GLANDS

CHAPTER 23   The Respiratory System

MULTIPLE CHOICE. Choose the one alternative that best completes the statement or answers the question.

1)   The lower respiratory system includes all of the following *except*:
   A) pharynx.   B) nose.
   C) trachea.   D) all of the above.

   Answer: C
   Type: MC   Page Ref: 681
   Topic: RESPIRATORY SYSTEM

2)   The floor of the internal nose:
   A) is also the hard palate.
   B) separates the internal nose from the external nose.
   C) is formed entirely by the palatine processes of the maxillae.
   D) is also called the nasal septum.

   Answer: A
   Type: MC   Page Ref: 681
   Topic: NOSE

3)   Which of the following does *not* form part of the nasal septum?
   A) cartilage   B) vomer
   C) ethmoid   D) inferior nasal conchae

   Answer: D
   Type: MC   Page Ref: 681
   Topic: NOSE

4)   Functions of the nose include:
   1. olfactory reception
   2. resonance of speech sounds
   3. filtration of incoming air
   4. warming of incoming air
   5. moistening of incoming air
   A) 1, 3, 4   B) 3, 4, 5   C) 1, 3, 4, 5   D) 1, 2, 3, 4, 5

   Answer: D
   Type: MC   Page Ref: 684
   Topic: NOSE

5)   Olfactory epithelium:
   A) is found on the superior nasal conchae and superior portion of the nasal septum.
   B) is the mucous membrane lining of the nasal cavity.
   C) lines the superior, middle, and inferior nasal meatuses.
   D) all of the above.

   Answer: A
   Type: MC   Page Ref: 684
   Topic: NOSE

6) Which of the following are openings into the nasopharynx?
   1. auditory tube
   2. fauces
   3. internal naris
   4. external naris
   5. nasolacrimal duct
   A) 1, 2, 3   B) 1, 3   C) 4, 5   D) 2, 3

   Answer: B
   Type: MC   Page Ref: 684
   Topic: PHARYNX

7) The epiglottis:
   A) consists of a leaf-shaped piece of hyaline cartilage.
   B) is attached posteriorly to the thyroid cartilage.
   C) serves to route food and liquids into the esophagus.
   D) all of the above.

   Answer: C
   Type: MC   Page Ref: 685
   Topic: LARYNX

8) Which of the following is a true statement concerning the larynx?
   A) The extrinsic muscles of the larynx connect the cartilages to each other.
   B) During swallowing, the larynx rises, causing the epiglottis to form a lid over the glottis.
   C) The lining of the larynx inferior to the vocal folds is stratified squamous epithelium.
   D) All of the above are true.

   Answer: B
   Type: MC   Page Ref: 685
   Topic: LARYNX

9) The _____ is an internal feature of the lower respiratory tract located at the bifurcation of the trachea into bronchi.
   A) carina   B) epiglottis   C) rimi vestibuli   D) cupula

   Answer: A
   Type: BI   Page Ref: 688
   Topic: TRACHEA

10) Which of the following is *not* true of the trachea?
    A) It is lined by a mucous membrane containing pseudostratified ciliated columnar epithelium, goblet cells, and basal cells.
    B) It is a tubular passageway, posterior to the esophagus.
    C) Its wall contains 16-20 C-shaped rings of hyaline cartilage.
    D) The outer layer of the wall of the trachea, the adventitia, is a layer of areolar connective tissue that joins the trachea to surrounding tissues.

    Answer: A
    Type: MC   Page Ref: 688
    Topic: TRACHEA

11) A bony landmark that identifies the point at which the trachea gives rise to bronchi is the _____.
A) suprasternal notch
B) level of first rib
C) superior body of the fifth thoracic vertebra
D) sternoclavicular joint

Answer: C
Type: BI   Page Ref: 688
Topic: BRONCHI

12) The _____ is/are the first portion(s) of the respiratory tract to enter lung tissue.
A) trachea   B) primary bronchi
C) secondary bronchi   D) tertiary bronchi

Answer: B
Type: BI   Page Ref: 691
Topic: BRONCHI

13) Which of the following do *not* contain cartilage?
1. tertiary bronchi
2. alveolar ducts
3. respiratory bronchioles
4. terminal bronchioles
A) 1, 2, 3, 4   B) 2, 3, 4   C) 2, 3   D) 2, 4

Answer: B
Type: MC   Page Ref: 691
Topic: BRONCHI

14) The base of a lung is:
A) the inferior surface.   B) concave.
C) just superior to the diaphragm. D) all of the above.

Answer: D
Type: MC   Page Ref: 692
Topic: LUNGS

15) Which of the following features of the lungs face(s) the heart?
1. costal surface
2. hilus
3. base
4. mediastinal surface
A) 1, 2   B) 3, 4   C) 2, 3   D) 1 only

Answer: C
Type: MC   Page Ref: 692
Topic: LUNGS

16) Compared to the left lung, the right lung:
A) is broader.   B) is shorter.
C) has one more lobe.   D) all of the above.

Answer: D
Type: MC   Page Ref: 692
Topic: LUNGS

17) The cardiac notch is a feature of the:
A) right lung.   B) left lung.   C) heart.   D) diaphragm.

Answer: B
Type: BI   Page Ref: 692
Topic: LUNGS

18) Each tertiary bronchus supplies a region of a lung called a/an _____.
A) segment   B) lobule   C) lobe   D) alveolar sac

Answer: A
Type: BI   Page Ref: 694
Topic: LUNGS

19) As a molecule of oxygen passes from alveolar air into the blood it passes through the layers of the alveolar capillary membrane in what order?
1. epithelial basement membrane
2. capillary basement membrane
3. alveolar epithelial wall
4. capillary endothelial wall
A) 3, 1, 2, 4   B) 4, 3, 2, 1   C) 1, 3, 2, 4   D) 3, 1, 4, 2

Answer: A
Type: MC   Page Ref: 694
Topic: LUNGS

20) Which of the following is *not* true of the diaphragm?
A) It is a striated (skeletal) muscle.
B) It receives innervation from cervical spinal nerves 3, 4, and 5 via the phrenic nerve.
C) In the relaxed state, the diaphragm is a flat, horizontal sheet.
D) When contracted, it decreases the vertical dimension of the abdominopelvic cavity.

Answer: C
Type: MC   Page Ref: 698
Topic: MECHANICS OF PULMONARY VENTILATION

21) Expansion and inflation of the lungs during inspiration is a result of:
A) contraction of the main inspiratory muscles.
B) surface tension between the parietal and visceral pleura.
C) atmospheric pressure.
D) all of the above.

Answer: D
Type: MC   Page Ref: 698
Topic: MECHANICS OF PULMONARY VENTILATION

22) Accessory muscles of inspiration are:
1. external intercostals
2. internal intercostals
3. sternocleidomastoid
4. scalenes
5. pectoralis major
6. pectoralis minor
A) 1, 3, 4, 5   B) 2, 3, 4, 6   C) 1, 4, 5, 6   D) 2, 3, 5

Answer: B
Type: MC   Page Ref: 698
Topic: MECHANICS OF PULMONARY VENTILATION

23) The principal muscles of inspiration are the _____ and _____.
A) diaphragm, internal intercostals
B) internal intercostals, external intercostals
C) diaphragm, sternocleidomastoid
D) diaphragm, external intercostals

Answer: D
Type: BI   Page Ref: 698
Topic: MECHANICS OF PULMONARY VENTILATION

24) Decrease in lung volume during normal expiration is mainly
due to:
A) relaxation of inspiratory muscles.   B) the inward pull of surface tension.
C) recoil of elastic fibers in the lungs.   D) all of the above.

Answer: D
Type: MC   Page Ref: 698
Topic: MECHANICS OF PULMONARY VENTILATION

25) CO_2 level in the blood is the primary factor that affects rate and depth of
ventilation. Increased CO_2 levels stimulate arterial and medullary chemoreceptors,
resulting in messages to the respiratory center to decrease the rate and depth of
ventilation.
A) Both statements are true.
B) Both statements are false.
C) The first statement is true; the second is false.
D) The second statement is true; the first is false.

Answer: C
Type: MC   Page Ref: 701
Topic: MECHANICS OF PULMONARY VENTILATION

TRUE/FALSE.   Write 'T' if the statement is true and 'F' if the statement is false.

1) The posterior portion of the nasal septum consists of the vomer, ethmoid, and
palatine bones. The anterior portion is almost entirely hyaline cartilage.

Answer: TRUE
Type: TF   Page Ref: 681
Topic: NOSE

2)   The cricoid cartilage is an important landmark in determining the site for a tracheostomy.
       Answer: TRUE
       Type: TF   Page Ref: 688
       Topic: LARYNX

3)   The true vocal cords are folds of mucous membrane called ventricular folds.
       Answer: FALSE
       Type: TF   Page Ref: 687
       Topic: LARYNX

4)   The vocal folds of adult females are usually thinner and vibrate more rapidly than the vocal folds of adult males.
       Answer: TRUE
       Type: TF   Page Ref: 687
       Topic: LARYNX

5)   The trachea is a tubular passageway for air, inferior to the larynx and anterior to the esophagus.
       Answer: TRUE
       Type: TF   Page Ref: 688
       Topic: TRACHEA

6)   An examination of the wall of the airways in the bronchial tree from bronchi to alveoli would reveal that cartilage decreases in quantity and disappears, while the amount of smooth muscle increases.
       Answer: TRUE
       Type: TF   Page Ref: 691
       Topic: BRONCHI

7)   The left primary bronchus is more vertical, shorter, and wider than the right primary bronchus.
       Answer: FALSE
       Type: TF   Page Ref: 690
       Topic: BRONCHI

8)   The horizontal fissure is superior to the middle lobe of the right lung.
       Answer: TRUE
       Type: TF   Page Ref: 694
       Topic: LUNGS

9)   The two types of pneumocytes in the alveolar wall are alveolar macrophages and alveolar septal cells.
       Answer: FALSE
       Type: TF   Page Ref: 694
       Topic: LUNGS

10) Pulmonary ventilation involves the physical movement of air between the alveoli and atmospheric air.

Answer: TRUE
Type: TF   Page Ref: 697
Topic: MECHANICS OF PULMONARY VENTILATION

11) The two portions of the respiratory center in the pons are the pneumotaxic and rhythmicity areas.

Answer: FALSE
Type: TF   Page Ref: 700
Topic: MECHANICS OF PULMONARY VENTILATION

ESSAY. Write your answer in the space provided or on a separate sheet of paper.

1) Describe the location of each of the three regions of the pharynx.

Answer: 1. Nasopharynx: posterior to nasal cavity from choanae to soft palate
   2. Oropharynx: posterior to oral cavity, from soft palate to hyoid
   3. Laryngopharynx: from hyoid bone to esophagus and larynx
Type: ES   Page Ref: 684
Topic: PHARYNX

2) Describe the lining tissue of the respiratory tract from the external nares to the beginning of the trachea. Relate the lining to the function of each area.

Answer: Vestibule: skin containing hairs (passage and filtering of air)
Vasal cavity, nasopharynx, and larynx: pseudostratified ciliated columnar epithelium with goblet cells (passage of air)
Oropharynx and laryngopharynx: stratified squamous epithelium (passage of air and food)
Larynx: pseudostratified ciliated columnar epithelium with goblet cells (passage of air)
Type: ES   Page Ref: 684-688
Topic: RESPIRATORY TRACT: Lining

3) Name and describe the nine pieces of cartilage of the larynx.

Answer: Answer should include a description of the three single cartilages: thyroid, epiglottic, cricoid; and the three pairs of cartilages: arytenoid, corniculate, and cuneiform.
Type: ES   Page Ref: 685-687
Topic: LARYNX

4) Describe the route travelled by a molecule of oxygen in inspired air from the time it enters the right external naris until it reaches an alveolus in the apex of the left lung.

Answer: Described route should include: vestibule, nasal cavity, internal nares, three regions of the pharynx, larynx, trachea, left primary bronchus, superior left secondary bronchus, left apical tertiary bronchus, bronchiole in apical segment, terminal bronchiole in lobule, respiratory bronchiole, alveolar duct, alveolus.
Type: ES   Page Ref: 681-694
Topic: RESPIRATORY SYSTEM

5)   List the cells found in the wall of an alveolus, stating the function of each one.

Answer: 1. Two types of pneumocytes make up the wall:
a) Type I alveolar cells (simple squamous epithelium) form a continuous lining which is the site of gas exchange.
b) Type II alveolar cells (septal cells) resemble cuboidal epithelium and secrete alveolar fluid containing surfactant, which helps prevent the collapse of alveoli during exhalation.
2. Alveolar macrophages (dust cells) are wandering macrophages that remove fine particles of dust.
3. Fibrobrasts produce elastic fibers (for recoil of lung tissue) and collagen fibers (for strength).

Type: ES   Page Ref: 694
Topic: LUNGS

SHORT ANSWER.   Write the word or phrase that best completes each statement or answers the question.

1)   The internal nose communicates posteriorly with the nasopharynx through two openings called _____.

Answer: internal nares, or choanae
Type: SA   Page Ref: 681
Topic: NOSE

2)   The anterior portion of the nasal cavity is called the _____.

Answer: vestibule
Type: SA   Page Ref: 681
Topic: NOSE

3)   The auditory tubes open into the _____ portion of the throat.

Answer: nasopharynx
Type: SA   Page Ref: 684
Topic: PHARYNX

4)   The laryngopharynx, or hypopharynx, directs food from the _____ to the _____.

Answer: oropharynx, esophagus
Type: SA   Page Ref: 684
Topic: PHARYNX

5)   Adam's apple is the common name for the _____ cartilage of the larynx.

Answer: thyroid
Type: SA   Page Ref: 685
Topic: LARYNX

6)   The laryngeal cartilages that, along with intrinsic pharyngeal muscles, move the vocal folds in speech production are the _____ cartilages.

Answer: arytenoid
Type: SA   Page Ref: 687
Topic: LARYNX

7) The open part of the C-shaped cartilages of the trachea face _____ (posteriorly or anteriorly).

Answer: posteriorly
Type: SA   Page Ref: 688
Topic: TRACHEA

8) Cilia in the upper respiratory tract beat so as to move mucus and trapped dust in a/an _____ (superior or inferior) direction.

Answer: inferior
Type: SA   Page Ref: 687
Topic: RESPIRATORY TRACT: Lining

9) Cilia in the lower respiratory tract beat so as to move mucus and trapped dust in a/an _____ (superior or inferior) direction.

Answer: superior
Type: SA   Page Ref: 687
Topic: RESPIRATORY TRACT: Lining

10) Sympathetic stimulation results in _____ (dilation or constriction) of bronchioles.

Answer: dilation
Type: SA   Page Ref: 691
Topic: BRONCHI

11) Type II alveolar cells secrete _____, which reduces the tendency of the alveoli to collapse during exhalation.

Answer: surfactant
Type: SA   Page Ref: 694
Topic: LUNGS

12) The three steps in the exchange of gases between the atmosphere, blood, and cells are _____, _____, and _____.

Answer: pulmonary ventilation, external (pulmonary) respiration, internal (tissue) respiration
Type: SA   Page Ref: 697
Topic: MECHANICS OF PULMONARY VENTILATION

13) The principal inspiratory muscles are the _____ and the _____.

Answer: diaphragm, external intercostals
Type: SA   Page Ref: 698
Topic: MECHANICS OF PULMONARY VENTILATION

14) The _____ area in the _____ of the brain controls the basic rate of pulmonary ventilation.

Answer: medullary rhythmicity, medulla oblongata
Type: SA   Page Ref: 700
Topic: MECHANICS OF PULMONARY VENTILATION

15) The two processes of pulmonary ventilation are _____ and _____.

Answer: inspiration, expiration
Type: SA   Page Ref: 697
Topic: MECHANICS OF PULMONARY VENTILATION

16) The _____ neurons of the respiratory center cause contraction of the diaphragm via signals sent through the phrenic nerve.

Answer: inspiratory
Type: SA   Page Ref: 700
Topic: MECHANICS OF PULMONARY VENTILATION

17) At about 4 weeks of fetal development, there is an outgrowth of endoderm of the foregut called the _____ bud, which develops into the future lungs and epithelial lining of parts of the respiratory tract.

Answer: laryngotracheal
Type: SA   Page Ref: 702
Topic: DEVELOPMENTAL ANATOMY

MATCHING. Choose the item in Column 2 that best matches each item in Column 1.

Match the laryngeal cartilages in Column 1 with the descriptions in Column 2.
1)   Column 1: thyroid
       Column 2: forms anterior wall of
                 larynx

       Answer: forms anterior wall of larynx
       Type: MA   Page Ref: 685
       Topic: LARYNX

2)   Column 1: cricoid
       Column 2: forms inferior portion of
                 wall of larynx

       Answer: forms inferior portion of wall of larynx
       Type: MA   Page Ref: 685
       Topic: LARYNX

3)   Column 1: arytenoid
       Column 2: attached to true vocal cords

       Answer: attached to true vocal cords
       Type: MA   Page Ref: 687
       Topic: LARYNX

4)   Column 1: corniculate
       Column 2: horn-shaped pieces of
                 elastic cartilage

       Answer: horn-shaped pieces of elastic cartilage
       Type: MA   Page Ref: 687
       Topic: LARYNX

5)   Column 1: epiglottis
       Column 2: movable flap of elastic
               cartilage

       Answer: movable flap of elastic cartilage
       Type: MA   Page Ref: 685
       Topic: LARYNX

Match the portions of the lower respiratory tract in Column 1 with the appropriate epithelial lining in Column 2.
   6)   Column 1: trachea
          Column 2: pseudostratified ciliated
                  columnar epithelium

          Answer: pseudostratified ciliated columnar epithelium
          Type: MA   Page Ref: 688
          Topic: TRACHEA

   7)   Column 1: primary bronchus
          Column 2: pseudostratified ciliated
                  columnar epithelium

          Answer: pseudostratified ciliated columnar epithelium
          Type: MA   Page Ref: 691
          Topic: BRONCHI

   8)   Column 1: terminal bronchioles
          Column 2: nonciliated simple cuboidal
                  epithelium

          Answer: nonciliated simple cuboidal epithelium
          Type: MA   Page Ref: 691
          Topic: BRONCHI

   9)   Column 1: alveolar ducts
          Column 2: simple squamous epithelium

          Answer: simple squamous epithelium
          Type: MA   Page Ref: 694
          Topic: LUNGS

   10)   Column 1: alveoli
          Column 2: two types of pneumocytes:
                  simple squamous epithelium
                  and secretory septal cells

          Answer: two types of pneumocytes: simple squamous epithelium and secretory septal
                  cells
          Type: MA   Page Ref: 694
          Topic: LUNGS

Match the branches of the bronchial tree in Column 1 with the descriptions in Column 2.

11) Column 1: secondary bronchus
Column 2: one in each lobe

Answer: one in each lobe
Type: MA   Page Ref: 691
Topic: BRONCHI

12) Column 1: tertiary bronchus
Column 2: ten in each lung

Answer: ten in each lung
Type: MA   Page Ref: 691
Topic: BRONCHI

13) Column 1: terminal bronchiole
Column 2: directs inspired air into a
                lobule

Answer: directs inspired air into a lobule
Type: MA   Page Ref: 691
Topic: BRONCHI

14) Column 1: left primary bronchus
Column 2: gives rise to two secondary
                bronchi
Foil: gives rise to three
                secondary bronchi

Answer: gives rise to two secondary bronchi
Type: MA   Page Ref: 691
Topic: BRONCHI

15) Column 1: alveolar duct
Column 2: connects alveoli to a
                respiratory bronchiole

Answer: connects alveoli to a respiratory bronchiole
Type: MA   Page Ref: 694
Topic: BRONCHI

CHAPTER 24   The Digestive System

MULTIPLE CHOICE.   Choose the one alternative that best completes the statement or answers the question.

1) The layer of connective tissue underlying the epithelium of the mucosa of the GI tract is called the_____.
    A) adventitia   B) lamina propria   C) submucosa   D) peritoneum

    Answer: B
    Type: BI   Page Ref: 708
    Topic: GENERAL HISTOLOGY OF THE GI TRACT

2) The peritoneal fold situated as a "fatty apron" anterior to the small intestine is the _____.

    A) mesentery   B) falciform ligament
    C) lesser omentum   D) greater omentum

    Answer: D
    Type: BI   Page Ref: 712, 713
    Topic: PERITONEUM

3) The structure attached to the soft palate that helps close off the nasopharynx during swallowing is the _____.
    A) fauces   B) uvula   C) epiglottis   D) tongue

    Answer: B
    Type: BI   Page Ref: 713
    Topic: MOUTH

4) Which of the following is an intrinsic muscle of the tongue?
    A) longitudinalis superior   B) hyoglossus
    C) genioglossus   D) styloglossus

    Answer: A
    Type: MC   Page Ref: 714
    Topic: TONGUE

5) The salivary glands whose ducts open on either side of the lingual frenulum are the:
    A) parotid glands.   B) sublingual glands.
    C) submandibular glands.   D) buccal glands.

    Answer: C
    Type: MC   Page Ref: 715
    Topic: SALIVARY GLANDS

6) The material of a tooth that attaches it to the periodontal ligament is _____.
    A) dentin   B) enamel   C) cementum   D) gingiva

    Answer: C
    Type: BI   Page Ref: 717
    Topic: TEETH

7) The functions of the esophagus include:
A) secretion of enzymes.
B) mixing of food and secretions.
C) absorption of H$_2$O and small nutrients.
D) none of the above.

Answer: D
Type: MC   Page Ref: 720
Topic: ESOPHAGUS

8) The stomach occupies the following abdominal region(s):
A) umbilical.   B) left hypochondriac.
C) epigastric.   D) all of the above.

Answer: D
Type: MC   Page Ref: 721
Topic: STOMACH

9) Place the following in the correct order, as found in the muscularis layer of the stomach, from outermost to innermost:
1. oblique layer
2. circular layer
3. longitudinal layer
A) 2, 1, 3   B) 3, 1, 2   C) 1, 2, 3   D) 3, 2, 1

Answer: D
Type: MC   Page Ref: 723
Topic: STOMACH

10) The head of the pancreas is located closest to the:
A) curve of the duodenum.   B) lesser curvature of the stomach.
C) inferior surface of the liver.   D) medial surface of the spleen.

Answer: A
Type: MC   Page Ref: 724
Topic: PANCREAS

11) A portal triad in the liver consists of branches of the:
A) common bile duct, and right and left hepatic ducts.
B) hepatic portal vein, hepatic artery, and bile duct.
C) hepatic portal vein, hepatic vein, and hepatic artery.
D) hepatic artery, central vein, and hepatic vein.

Answer: B
Type: MC   Page Ref: 726
Topic: LIVER

12) The functional unit of the liver is a:
A) lobe.   B) lobule.
C) sinusoid.   D) bile canaliculus.

Answer: B
Type: MC   Page Ref: 728
Topic: LIVER

13) Bile is produced by liver cells called:
A) stellate reticuloendothelial cells.
B) phagocytes.
C) hepatocytes.
D) endothelial cells.

Answer: C
Type: MC   Page Ref: 728
Topic: LIVER

14) Bile helps accomplish which of the following?
1. chemical breakdown of protein
2. emulsification
3. absorption of fats
4. maintenance of an alkaline pH in the duodenum
A) 1, 2, 3, 4   B) 1, 2, 3   C) 2, 3, 4   D) 1, 3, 4

Answer: C
Type: MC   Page Ref: 729
Topic: LIVER

15) The mucosa of the gallbladder:
A) secretes bile.
B) stores bile.
C) removes water and ions from bile.
D) consists of stratified columnar epithelium.

Answer: C
Type: MC   Page Ref: 729
Topic: GALLBLADDER

16) Peristaltic contractions:
A) occur in the stomach, esophagus, small intestine, and large intestine.
B) are primarily responsible for mixing the food with the digestive juices.
C) are of similar strength throughout the digestive tract.
D) all of the above.

Answer: A
Type: MC   Page Ref: 720,724,732,737
Topic: SMALL INTESTINE

17) Which of the following is an anatomical feature of the small intestine that serves to increase the surface area for digestion and absorption?
A) lacteals
B) microvilli
C) rugae
D) all of the above

Answer: B
Type: MC   Page Ref: 730
Topic: SMALL INTESTINE

18) Which of the following is *not* a movement of the large intestine?
A) peristalsis
B) mass peristalsis
C) segmentation
D) haustral churning

Answer: C
Type: MC   Page Ref: 732, 737
Topic: LARGE INTESTINE

19) Contraction of which of the following aid in defecation?
1. external anal sphincter
2. circular colon muscles
3. longitudinal rectal muscles
4. internal anal sphincter
5. diaphragm
6. abdominal muscles

A) 2, 3, 5, 6   B) 1, 2, 3, 4   C) 1, 2, 4   D) 3, 5, 6

Answer: D
Type: MC   Page Ref: 737, 738
Topic: LARGE INTESTINE

TRUE/FALSE. Write 'T' if the statement is true and 'F' if the statement is false.

1) Enzymatic breakdown of food is a form of mechanical digestion.

Answer: FALSE
Type: TF   Page Ref: 708
Topic: DIGESTIVE PROCESSES

2) In the wall of the GI tract, the muscularis mucosae and the muscularis each consist of two layers of smooth muscle cells, an outer longitudinal and an inner circular layer.

Answer: TRUE
Type: TF   Page Ref: 708, 709
Topic: GENERAL HISTOLOGY OF THE GI TRACT

3) The palatine tonsils are located in the uvula, at the posterior border of the soft palate.

Answer: FALSE
Type: TF   Page Ref: 713
Topic: MOUTH

4) The tongue is attached to the floor of the oral cavity by a fold of mucous membrane called the labial frenulum.

Answer: FALSE
Type: TF   Page Ref: 713
Topic: MOUTH

5) The secretory activity of the salivary glands is controlled via vasoconstriction and vasodilation of their blood vessels.

Answer: FALSE
Type: TF   Page Ref: 717
Topic: SALIVARY GLANDS

6) In the permanent dentition, the teeth that have two cusps and two roots are the upper premolars (bicuspids).

Answer: TRUE
Type: TF   Page Ref: 718
Topic: TEETH

7)   The muscularis layer of the esophagus contains both skeletal and smooth muscle throughout its length.
      Answer: FALSE
      Type: TF   Page Ref: 720
      Topic: ESOPHAGUS

8)   The greater curvature of the stomach is closer to the liver than is the lesser curvature.
      Answer: FALSE
      Type: TF   Page Ref: 721
      Topic: STOMACH

9)   As a result of the action of pancreatic enzymes, the following are digested in the lumen of the small intestine: carbohydrates, proteins, triglycerides, nucleic acids.
      Answer: TRUE
      Type: TF   Page Ref: 726
      Topic: PANCREAS

10)   Most pancreatic tissue is exocrine in function and is arranged in cell clusters called acini.
      Answer: TRUE
      Type: TF   Page Ref: 726
      Topic: PANCREAS

11)   The pancreas is the largest and heaviest gland of the body.
      Answer: FALSE
      Type: TF   Page Ref: 726
      Topic: LIVER

12)   The hepatic veins deliver nutrient-rich blood to the sinusoids of the liver.
      Answer: FALSE
      Type: TF   Page Ref: 726
      Topic: LIVER

13)   Mucus in the intestine is secreted by goblet cells in the mucosa and by duodenal glands in the submucosa.
      Answer: TRUE
      Type: TF   Page Ref: 732
      Topic: SMALL INTESTINE

14)   Pouches of the large intestine that give it a puckered appearance are called epiploic appendages.
      Answer: FALSE
      Type: TF   Page Ref: 737
      Topic: LARGE INTESTINE

15) The epithelial lining and glands of most of the GI tract are derived from endoderm; the smooth muscle and connective tissue develop from mesoderm.

Answer: TRUE
Type: TF   Page Ref: 739
Topic: DEVELOPMENTAL ANATOMY

ESSAY.   Write your answer in the space provided or on a separate sheet of paper.

1) Describe the structure of a tooth and illustrate your answer with a labeled diagram.

Answer: Diagram and description should contain: crown, root, neck, dentin, enamel, pulp cavity, pulp, root canal, apical foramen, cementum, periodontal ligament. Refer to Fig. 24.7.
Type: ES   Page Ref: 717
Topic: TEETH

2) Describe the teeth of the permanent dentition as found on one–half of the jaw, in order from anterior to posterior.

Answer: Refer to Fig. 24.8b.
Type: ES   Page Ref: 717–718
Topic: TEETH

3) Describe the gross anatomy of the stomach. Illustrate your answer with a labeled diagram.

Answer: Description and diagram should include the four regions of the stomach, the two regions of the pylorus and sphincter, rugae, and greater and lesser curvatures. See Fig. 24.10.
Type: ES   Page Ref: 721
Topic: STOMACH

4) Trace the route most commonly traveled by pancreatic juice, from its production until it is in the lumen of the intestine.

Answer: Pancreatic juice may travel through the following: acini, small ducts, pancreatic duct, hepatopancreatic ampulla, major duodenal papilla.
Type: ES   Page Ref: 724
Topic: PANCREAS

5) Describe the route traveled by bile from the time it is secreted until it enters the duodenum.

Answer: Bile travels through bile canaliculi (bile capillaries), bile ducts, right or left hepatic duct, common hepatic duct, cystic duct, gall bladder, cystic duct, common bile duct, hepatopancreatic ampulla.
Type: ES   Page Ref: 728–729
Topic: LIVER

6) Describe the route traveled by chyme from the time it leaves the ileum until it is expelled as feces from the body. As you mention the regions of the large intestine, state which are retroperitoneal.

Answer: The following terms should be included: ileocecal valve, cecum, ascending colon (retroperitoneal), r. colic flexure, transverse colon, l. colic flexure, descending colon (retroperitoneal), sigmoid colon, rectum, anal canal, anus (internal involuntary sphincter, external voluntary sphincter).
Type: ES   Page Ref: 734-736
Topic: LARGE INTESTINE

SHORT ANSWER. Write the word or phrase that best completes each statement or answers the question.

1) The _____ plexus controls GI tract motility; the _____ plexus controls GI tract secretions.

Answer: myenteric, submucosal
Type: SA   Page Ref: 710, 711
Topic: GENERAL HISTOLOGY OF THE GI TRACT

2) The largest serous membrane of the body is the _____.

Answer: peritoneum
Type: SA   Page Ref: 711
Topic: PERITONEUM

3) The only digestive organ attached to the anterior abdominal wall is the _____.

Answer: liver
Type: SA   Page Ref: 712
Topic: PERITONEUM

4) The _____ of the oral cavity is a space that extends externally from the teeth and gums to the cheek and lips.

Answer: vestibule
Type: SA   Page Ref: 713
Topic: MOUTH

5) The _____ is the passageway between the oral cavity and the oropharynx.

Answer: fauces
Type: SA   Page Ref: 713
Topic: MOUTH

6) The papillae of the tongue that contain taste buds are _____ and _____.
Answer: fungiform, circumvallate
Type: SA   Page Ref: 714
Topic: TONGUE

7) The calcified connective tissue that forms most of a tooth, giving it shape and rigidity, is called _____.
Answer: dentin
Type: SA   Page Ref: 717
Topic: TEETH

8) The opening in the diaphragm through which the esophagus passes is the _____.

Answer: esophageal hiatus
Type: SA   Page Ref: 720
Topic: ESOPHAGUS

9) The _____ is the inferior portion of the stomach that connects with the duodenum through the _____ sphincter.

Answer: pylorus, pyloric
Type: SA   Page Ref: 721
Topic: STOMACH

10) The hepatopancreatic ampulla is formed by the merging of the _____ duct and the _____ duct.

Answer: common bile, pancreatic
Type: SA   Page Ref: 724
Topic: PANCREAS

11) The four lobes of the liver are the _____, _____, _____, and _____.

Answer: right, left, quadrate, caudate
Type: SA   Page Ref: 726
Topic: LIVER

12) The organ of the digestive system that receives both arterial (oxygenated) and venous (deoxygenated) blood is the _____.

Answer: liver
Type: SA   Page Ref: 726
Topic: LIVER

13) Rugae are folds in the mucous membrane lining of two digestive organs: the _____ and the _____.

Answer: stomach, gallbladder
Type: SA   Page Ref: 721, 729
Topic: GALLBLADDER

14) Components of MALT, lymphatic nodules are located in the wall of the _____ of the small intestine.

Answer: ileum (third segment)
Type: SA   Page Ref: 732
Topic: SMALL INTESTINE

15) Movement in the small intestine that is primarily responsible for mixing chyme with intestinal juice and that also facilitates absorption is called _____.

Answer: segmentation
Type: SA   Page Ref: 732
Topic: SMALL INTESTINE

16) The longitudinal smooth muscle of the colon is arranged in three bands called _____.

Answer: taeniae coli
Type: SA   Page Ref: 736
Topic: LARGE INTESTINE

17)   _____ of the large intestine is/are responsible for the final stage of digestion.

Answer: bacteria
Type: SA   Page Ref: 737
Topic: LARGE INTESTINE

18)   Oxygenated blood is supplied to the colon via branches of the abdominal aorta, the
_____ and the _____.

Answer: superior mesenteric artery, inferior mesenteric artery
Type: SA   Page Ref: 736
Topic: LARGE INTESTINE

19)   Deoxygenated blood from the colon flows through superior and inferior mesenteric
veins to the _____.

Answer: hepatic portal vein or liver
Type: SA   Page Ref: 736
Topic: LARGE INTESTINE

MATCHING.   Choose the item in Column 2 that best matches each item in Column 1.

Match the layers of the wall of the GI tract in Column 1 with their descriptions in Column 2.
1)   Column 1: lining epithelium
       Column 2: innermost layer

Answer: innermost layer
Type: MA   Page Ref: 708
Topic: GENERAL HISTOLOGY OF THE GI TRACT

2)   Column 1: lamina propria
       Column 2: areolar tissue that contains
                 most of the mucosa-
                 associated lymphoid tissue

Answer: areolar tissue that contains most of the mucosa-associated lymphoid tissue
Type: MA   Page Ref: 708
Topic: GENERAL HISTOLOGY OF THE GI TRACT

3)   Column 1: submucosa
       Column 2: areolar tissue that binds
                 the mucosa to the
                 muscularis

Answer: areolar tissue that binds the mucosa to the muscularis
Type: MA   Page Ref: 710
Topic: GENERAL HISTOLOGY OF THE GI TRACT

4)   Column 1: muscularis
       Column 2: may contain skeletal muscle
                 in the mouth, pharynx, and
                 upper esophagus

Answer: may contain skeletal muscle in the mouth, pharynx, and upper esophagus
Type: MA   Page Ref: 710
Topic: GENERAL HISTOLOGY OF THE GI TRACT

5)   Column 1: serosa
        Column 2: called visceral peritoneum
                  in abdominal region

        Answer: called visceral peritoneum in abdominal region
        Type: MA   Page Ref: 711
        Topic: GENERAL HISTOLOGY OF THE GI TRACT

6)   Column 1: muscularis mucosae
        Column 2: smooth muscle layer
                  between lamina propria and
                  submucosa layers

        Answer: smooth muscle layer between lamina propria and submucosa layers
        Type: MA   Page Ref: 708
        Topic: GENERAL HISTOLOGY OF THE GI TRACT

Match the cells of the gastric glands in Column 1 with their secretions in Column 2.
7)   Column 1: chief cells
        Column 2: pepsinogen and lipase
        Foil: lipase and hydrochloric acid

        Answer: pepsinogen and lipase
        Type: MA   Page Ref: 723
        Topic: STOMACH

8)   Column 1: parietal cells
        Column 2: hydrochloric acid and
                  intrinsic factor

        Answer: hydrochloric acid and intrinsic factor
        Type: MA   Page Ref: 723
        Topic: STOMACH

9)   Column 1: mucous cells
        Column 2: mucus
        Foil: mucus and pepsinogen

        Answer: mucus
        Type: MA   Page Ref: 723
        Topic: STOMACH

10)   Column 1: G cells
        Column 2: gastrin

        Answer: gastrin
        Type: MA   Page Ref: 723
        Topic: STOMACH

Match the functions of the liver in Column 1 with the most appropriate words in Column 2.
11)   Column 1: excretes
        Column 2: bilirubin

        Answer: bilirubin
        Type: MA   Page Ref: 729
        Topic: LIVER

12) Column 1: stores
Column 2: iron

Answer: iron
Type: MA   Page Ref: 729
Topic: LIVER

13) Column 1: breaks this(these) down to
produce acetyl coenzyme A
Column 2: fatty acids

Answer: fatty acids
Type: MA   Page Ref: 729
Topic: LIVER

14) Column 1: synthesizes
Column 2: albumin

Answer: albumin
Type: MA   Page Ref: 729
Topic: LIVER

15) Column 1: phagocytizes
Column 2: bacteria

Answer: bacteria
Type: MA   Page Ref: 729
Topic: LIVER

Match the regions of the peritoneum in Column 1 with their descriptions in Column 2.

16) Column 1: mesentery
Column 2: holds small intestine loosely
in place and binds it to
the posterior abdominal
wall

Answer: holds small intestine loosely in place and binds it to the posterior abdominal
wall
Type: MA   Page Ref: 712
Topic: PERITONEUM

17) Column 1: mesocolon
Column 2: holds large intestine loosely
in place and binds it to
the posterior abdominal
wall

Answer: holds large intestine loosely in place and binds it to the posterior abdominal
wall
Type: MA   Page Ref: 712
Topic: PERITONEUM

18)   Column 1: mesocolon
         Column 2: location of nerves and
                  blood vessels that serve the
                  large intestine

         Answer: location of nerves and blood vessels that serve the large intestine
         Type: MA   Page Ref: 712
         Topic: PERITONEUM

19)   Column 1: falciform ligament
         Column 2: attaches liver to the
                  diaphragm and anterior
                  abdominal wall

         Answer: attaches liver to the diaphragm and anterior abdominal wall
         Type: MA   Page Ref: 712
         Topic: PERITONEUM

20)   Column 1: lesser omentum
         Column 2: fold that runs from the
                  stomach and duodenum to
                  the liver

         Answer: fold that runs from the stomach and duodenum to the liver
         Type: MA   Page Ref: 712
         Topic: PERITONEUM

21)   Column 1: greater omentum
         Column 2: a four-layered sheet that is
                  suspended from the
                  stomach, duodenum, and
                  transverse colon

         Answer: a four-layered sheet that is suspended from the stomach, duodenum, and
                  transverse colon
         Type: MA   Page Ref: 712
         Topic: PERITONEUM

22)   Column 1: greater omentum
         Column 2: contains large quantity of
                  adipose tissue and many
                  lymph nodes

         Answer: contains large quantity of adipose tissue and many lymph nodes
         Type: MA   Page Ref: 712
         Topic: PERITONEUM

CHAPTER 25   The Urinary System

MULTIPLE CHOICE. Choose the one alternative that best completes the statement or answers the question.

1) The urinary system contains some retroperitoneal organs. They include:
   A) kidneys only.
   B) urinary bladder.
   C) ureters only.
   D) all of the above.

   Answer: D
   Type: MC   Page Ref: 749, 761, 762
   Topic: KIDNEYS

2) Which of the following is *not* true regarding the location of the kidneys?
   A) They are partially protected by the floating ribs.
   B) The left kidney is slightly lower than the right.
   C) They are located posterior to the parietal peritoneum of the posterior abdominal wall.
   D) They are situated between the levels of the last thoracic and third lumbar vertebrae.

   Answer: B
   Type: MC   Page Ref: 749
   Topic: KIDNEYS

3) Three layers of tissue surround each kidney. They are, in order from innermost to outermost:
   1. renal capsule
   2. visceral peritoneum (serosa)
   3. adipose capsule
   4. renal fascia
   A) 1, 3, 4   B) 2, 3, 4   C) 1, 3, 2   D) 4, 3, 1

   Answer: A
   Type: MC   Page Ref: 749-751
   Topic: KIDNEYS

4) Trace the route of an oxygenated red blood cell from the time it passes into the kidney in a renal artery until it enters a venule as a deoxygenated red blood cell by placing the following vessels in their correct order:
   1. segmental artery
   2. arcuate artery
   3. interlobar artery
   4. peritubular capillary
   5. afferent arteriole
   6. peritubular venule
   7. efferent arteriole
   8. interlobular artery
   A) 1, 3, 2, 8, 5, 9, 7, 4, 6   B) 3, 2, 1, 8, 4, 7, 9, 5, 6
   C) 1, 8, 2, 3, 7, 9, 5, 4, 6   D) 1, 2, 3, 8, 5, 9, 7, 6, 4

   Answer: A
   Type: MC   Page Ref: 752-754
   Topic: KIDNEYS

5)	A nephron consists of two portions:
	A) renal corpuscle and renal tubule.
	B) glomerulus and glomerular capsule.
	C) proximal and distal convoluted tubules.
	D) glomerulus and collecting duct.

	Answer: A
	Type: MC Page Ref: 754
	Topic: NEPHRON

6)	Which of the following structures is *not* a portion of a nephron?
	A) vasa recta	B) loop of Henle
	C) distal convoluted tubule	D) proximal convoluted tubule

	Answer: A
	Type: MC Page Ref: 754
	Topic: NEPHRON

7)	The basement membrane layer of the endothelial–capsular membrane:
	A) is a single layer of endothelial cells.
	B) contains cells called podocytes that prevent blood cells from entering the filtrate.
	C) is a layer of extracellular material between the parietal and visceral layers of the glomerular capsule.
	D) none of the above.

	Answer: D
	Type: MC Page Ref: 758
	Topic: NEPHRON

8)	Cortical nephrons lack one portion of the renal tubule that is found in juxtamedullary nephrons, the:
	A) thick segment of the ascending limb of the loop of Henle.
	B) thin segment of the ascending limb of the loop of Henle.
	C) proximal convoluted tubule.
	D) distal convoluted tubule.

	Answer: B
	Type: MC Page Ref: 754
	Topic: NEPHRON

9)	The two portions of a nephron that commonly contribute to the juxtaglomerular apparatus are the:
	A) glomerulus and distal convoluted tubule.
	B) descending limb of loop of Henle and efferent arteriole.
	C) distal portion of ascending limb of loop of Henle and afferent arteriole.
	D) glomerulus and collecting duct.

	Answer: C
	Type: MC Page Ref: 758
	Topic: JUXTAGLOMERULAR APPARATUS

10) Juxtaglomerular cells:
   A) monitor the salt concentration of the fluid in the renal tubule.
   B) are found in the macula densa.
   C) are modified smooth muscle cells that secrete renin, which helps regulate blood pressure.
   D) all of the above.

Answer: C
Type: MC   Page Ref: 758
Topic: JUXTAGLOMERULAR APPARATUS

11) Normally, urine is prevented from backing up into the ureters from a full bladder due to:
   A) gravity.
   B) hydrostatic pressure from the renal pelvis.
   C) sphincters at the junctions of the ureters and bladder.
   D) hydrostatic pressure from the bladder that compresses the ureteral openings.

Answer: D
Type: MC   Page Ref: 761
Topic: URETERS

12) The urinary bladder is directly posterior to the pubic symphysis in males. In females, the urinary bladder is posterior to the vagina.
   A) Both statements are true.
   B) Both statements are false.
   C) The first statement is true; the second is false.
   D) The second statement is true; the first is false.

Answer: C
Type: MC   Page Ref: 762
Topic: URINARY BLADDER

13) The internal urethral sphincter is derived from circular smooth muscle. The external urethral sphincter is derived from voluntary skeletal muscle.
   A) Both statements are true.
   B) Both statements are false.
   C) The first statement is true; the second is false.
   D) The second statement is true; the first is false.

Answer: A
Type: MC   Page Ref: 762
Topic: URINARY BLADDER

14) The lining of the male urethra contains:
   A) transitional epithelium.   B) pseudostratified epithelium.
   C) stratified squamous epithelium.   D) all of the above.

Answer: D
Type: MC   Page Ref: 764
Topic: URETHRA

15) The shortest portion of the male urethra is the _____.
   A) spongy urethra   B) membranous urethra
   C) prostatic urethra   D) penile urethra
   Answer: B
   Type: BI   Page Ref: 764
   Topic: URETHRA

TRUE/FALSE.   Write 'T' if the statement is true and 'F' if the statement is false.

1) The urinary system consists of two kidneys, two urethras, one urinary bladder, and one ureter.
   Answer: FALSE
   Type: TF   Page Ref: 749
   Topic: URETERS

2) A renal papilla is located at the base of each renal pyramid.
   Answer: FALSE
   Type: TF   Page Ref: 751
   Topic: KIDNEYS

3) The renal corpuscle, proximal convoluted tubule, and distal convoluted tubule of a nephron are located in the renal cortex.
   Answer: TRUE
   Type: TF   Page Ref: 754
   Topic: NEPHRON

4) The kidneys contain more cortical nephrons than juxtamedullary nephrons.
   Answer: TRUE
   Type: TF   Page Ref: 754
   Topic: NEPHRON

5) Intercalated cells of the collecting ducts secrete excess H^+ into the urine.
   Answer: TRUE
   Type: TF   Page Ref: 758
   Topic: NEPHRON

6) The salt concentration of urine is monitored by specialized cells of the renal tubule just as the urine enters the distal convoluted tubule.
   Answer: TRUE
   Type: TF   Page Ref: 758
   Topic: JUXTAGLOMERULAR APPARATUS

7) At rest, approximately 50% of cardiac output is directed to the kidneys.
   Answer: FALSE
   Type: TF   Page Ref: 752
   Topic: KIDNEYS

8) The process of tubular reabsorption rids the body of some materials such as hydrogen ions and ammonium ions and some drugs such as penicillin.

Answer: FALSE
Type: TF   Page Ref: 761
Topic: URINE FORMATION

9) The distal region of a ureter has two layers of smooth muscle in the muscularis.

Answer: FALSE
Type: TF   Page Ref: 761
Topic: URETERS

10) There are three openings into the urinary bladder: a pair of ureteral openings and an internal urethral orifice.

Answer: TRUE
Type: TF   Page Ref: 762
Topic: URINARY BLADDER

11) The female urethra opens between the clitoris and the vaginal opening.

Answer: TRUE
Type: TF   Page Ref: 764
Topic: URETHRA

12) The kidneys begin excreting urine by the third month of development, thus contributing most of the volume of the amniotic fluid.

Answer: FALSE
Type: TF   Page Ref: 765
Topic: DEVELOPMENTAL ANATOMY

ESSAY. Write your answer in the space provided or on a separate sheet of paper.

1) List the functions of the urinary system.

Answer: Three main kidney functions are on pp. 748. In addition, the ureters and urethra transport urine; the urinary bladder is a site of temporary storage.
Type: ES   Page Ref: 748
Topic: URINARY SYSTEM: Functions

2) Make a diagram of a coronal section of a kidney. Label the renal capsule, cortex, medulla, renal pyramids, renal column, renal papilla, minor calyx, and renal pelvis.

Answer: See Fig. 25.4(a).
Type: ES   Page Ref: 751
Topic: KIDNEYS

3) Describe the locations of a cortical and a juxtamedullary nephron.

Answer: 1. Cortical nephron: renal corpuscle is in the outer cortex; loop of Henle is short and penetrates the outer region of the medulla.
2. Juxtamedullary nephron: renal corpuscle is deep in the cortex near the medulla; the long loop of Henle penetrates deep into a pyramid almost to the apex.
Type: ES   Page Ref: 754
Topic: NEPHRON

4) Describe the layers through which a substance will pass when moving from glomerular blood into filtrate of the capsular space.

Answer: 1. Endothelial fenestrations of the glomerulus: large pores that prevent blood cells from entering the filtrate
2. Basement membrane of the glomerulus: glycoprotein matrix and fibrils between the endothelium and the visceral layer of the glomerular capsule; prevents passage of large proteins into the filtrate
3. Slit membranes between pedicels of podocytes (cells of the visceral layer of the glomerular capsule): the slit membrane prevents the filtration of medium-sized proteins

Type: ES   Page Ref: 758
Topic: NEPHRON

5) Urine formation involves three processes. Define them and name the regions of the nephron and blood supply involved in each process.

Answer: Include the proper use of the following: glomerular filtration, endothelial-capsular membrane, filtrate, glomerular blood pressure, tubular reabsorption, renal tubules, vasa recta, peritubular capillaries, tubular secretion, as on pp. 759-760.

Type: ES   Page Ref: 759-760
Topic: NEPHRON

6) Describe the three coats that make up the wall of the urinary bladder, from innermost to outermost.

Answer: 1. Mucosa of transitional epithelium and lamina propria
2. Detrusor muscle of inner longitudinal, middle circular, and outer longitudinal smooth muscle
3. Peritoneum on the superior surface, fibrous connective tissue coat on the remainder

Type: ES   Page Ref: 762
Topic: URINARY BLADDER

SHORT ANSWER. Write the word or phrase that best completes each statement or answers the question.

1) The indentation on the medial border of a kidney where blood and lymphatic vessels, nerves, and the ureter are attached is called the _____.

Answer: hilus
Type: SA   Page Ref: 749
Topic: KIDNEYS

2) As urine leaves the collecting duct of a pyramid, it enters a/an _____.

Answer: minor calyx
Type: SA   Page Ref: 752
Topic: KIDNEYS

3) Three functions of a nephron that result in urine formation are _____, _____, and _____.

Answer: filtration, reabsorption, secretion
Type: SA   Page Ref: 754
Topic: NEPHRON

4) The renal corpuscle consists of two parts: the _____ and the _____.

Answer: glomerulus, glomerular (Bowman's) capsule
Type: SA   Page Ref: 754
Topic: NEPHRON

5) Lining the collecting ducts are epithelial cells called _____ cells that are sensitive to ADH.

Answer: principal
Type: SA   Page Ref: 758
Topic: NEPHRON

6) The loop of Henle of juxtamedullary nephrons is accompanied by a blood vessel loop called the _____.

Answer: vasa recta
Type: SA   Page Ref: 754
Topic: NEPHRON

7) Each ureter is an extension of the _____ of the kidney.

Answer: pelvis
Type: SA   Page Ref: 761
Topic: URETERS

8) The inner coat of a ureter is protected from acidic hypertonic urine by _____.

Answer: mucus
Type: SA   Page Ref: 761
Topic: URETERS

9) The smooth triangular region in the floor of the urinary bladder is called the _____.

Answer: trigone
Type: SA   Page Ref: 762
Topic: URINARY BLADDER

10) You have now studied three saclike organs that have rugae on the inner lining: the _____, _____, and _____.

Answer: stomach, gall bladder, urinary bladder
Type: SA   Page Ref: 721, 729, 762
Topic: URINARY BLADDER

11) The regions of the male urethra, in order from proximal to distal, are _____, _____, and _____.

Answer: prostatic, membranous, spongy (penile)
Type: SA   Page Ref: 764
Topic: URETHRA

MATCHING. Choose the item in Column 2 that best matches each item in Column 1.

Trace the route of a water molecule from the time it is in the arteriole on its way into the renal corpuscle until it is released from the apex of a renal pyramid. Match the locations in Column 2 with the correct number, in sequence, in Column 1.

1)   Column 1: 1.
        Column 2: afferent arteriole
        Foil: efferent arteriole

        Answer: afferent arteriole
        Type: MA   Page Ref: 754
        Topic: NEPHRON

2)   Column 1: 2.
        Column 2: glomerulus

        Answer: glomerulus
        Type: MA   Page Ref: 754
        Topic: NEPHRON

3)   Column 1: 3.
        Column 2: capsular space of
                  glomerular capsule

        Answer: capsular space of glomerular capsule
        Type: MA   Page Ref: 754
        Topic: NEPHRON

4)   Column 1: 4.
        Column 2: proximal convoluted tubule

        Answer: proximal convoluted tubule
        Type: MA   Page Ref: 754
        Topic: NEPHRON

5)   Column 1: 5.
        Column 2: descending limb of the loop
                  of Henle

        Answer: descending limb of the loop of Henle
        Type: MA   Page Ref: 754
        Topic: NEPHRON

6)   Column 1: 6.
        Column 2: ascending limb of the loop
                  of Henle

        Answer: ascending limb of the loop of Henle
        Type: MA   Page Ref: 754
        Topic: NEPHRON

7)   Column 1: 7.
        Column 2: distal convoluted tubule

        Answer: distal convoluted tubule
        Type: MA   Page Ref: 754
        Topic: NEPHRON

8) Column 1: 8.
Column 2: collecting duct

Answer: collecting duct
Type: MA   Page Ref: 754
Topic: NEPHRON

9) Column 1: 9.
Column 2: papillary duct

Answer: papillary duct
Type: MA   Page Ref: 754
Topic: NEPHRON

10) Column 1: 10.
Column 2: minor calyx

Answer: minor calyx
Type: MA   Page Ref: 754
Topic: NEPHRON

Match the regions of the urinary system in Column 1 with the corresponding type of lining epithelial cells in Column 2.

11) Column 1: visceral layer of glomerular
capsule
Column 2: podocytes

Answer: podocytes
Type: MA   Page Ref: 758
Topic: EPITHELIAL LINING

12) Column 1: proximal convoluted tubule
Column 2: simple cuboidal with
microvilli

Answer: simple cuboidal with microvilli
Type: MA   Page Ref: 758
Topic: EPITHELIAL LINING

13) Column 1: descending limb of loop of
Henle
Column 2: simple squamous

Answer: simple squamous
Type: MA   Page Ref: 758
Topic: EPITHELIAL LINING

14) Column 1: thick ascending limb of
loop of Henle
Column 2: simple cuboidal to low
columnar

Answer: simple cuboidal to low columnar
Type: MA   Page Ref: 758
Topic: EPITHELIAL LINING

15) Column 1: distal convoluted tubule
Column 2: mixture of principal cells
and intercalated cells

Answer: mixture of principal cells and intercalated cells
Type: MA   Page Ref: 758
Topic: EPITHELIAL LINING

16) Column 1: papillary ducts
Column 2: simple columnar

Answer: simple columnar
Type: MA   Page Ref: 758
Topic: EPITHELIAL LINING

17) Column 1: ureter
Column 2: transitional epithelium

Answer: transitional epithelium
Type: MA   Page Ref: 761
Topic: EPITHELIAL LINING

CHAPTER 26   The Reproductive Systems

MULTIPLE CHOICE. Choose the one alternative that best completes the statement or answers the question.

1)   The dartos muscle:
A) causes wrinkling of the skin of the scrotum.
B) is found in the scrotal septum.
C) relaxes in response to warmth, contracts in response to cold.
D) all of the above.

Answer: C
Type: MC   Page Ref: 771
Topic: SCROTUM

2)   Descent of the testes into the scrotum normally occurs:
A) near the end of the seventh month of fetal development.
B) in the ninth month of fetal development.
C) shortly after birth.
D) at puberty.

Answer: A
Type: MC   Page Ref: 771
Topic: TESTES

3)   Which of the following pairs of terms is most closely matched?
A) tunica vaginalis; testes   B) tunica albuginea; uterus
C) germinal epithelium; spermatic cord   D) corpus luteum; penis

Answer: A
Type: MC   Page Ref: 772
Topic: TESTES

4)   The blood-testis barrier:
A) consists of tight junctions between interstitial cells of Leydig.
B) is the basement membrane of the seminiferous tubules.
C) prevents sperm antigens from getting into the bloodstream.
D) all of the above.

Answer: C
Type: MC   Page Ref: 773
Topic: TESTES

5)   Spermatogenesis and oogenesis are the *only* processes in the body where _____ occurs.
A) mitosis   B) meiosis   C) replication   D) mutation

Answer: B
Type: BI   Page Ref: 776
Topic: SPERMATOGENESIS

6) Which of the following applies to the first stage of meiosis in the male?
   A) reduction division of a primary spermatocyte
   B) equatorial division of a primary spermatocyte
   C) reduction division of a secondary spermatocyte
   D) equatorial division of a secondary spermatocyte

   Answer: A
   Type: BI   Page Ref: 776
   Topic: SPERMATOGENESIS

7) Meiosis results in:
   A) two new cells with 23 chromosomes each.
   B) four new cells with 23 chromosomes each.
   C) two new cells with 46 chromosomes each.
   D) four new cells with 46 chromosomes each.

   Answer: B
   Type: BI   Page Ref: 776
   Topic: SPERMATOGENESIS

8) The male reproductive duct that is six meters long but is tightly coiled into a 3.5-cm distance is the _____.
   A) rete testis   B) ductus epididymis
   C) ductus deferens   D) ejaculatory duct

   Answer: B
   Type: BI   Page Ref: 778
   Topic: DUCTS OF THE MALE REPRODUCTIVE TRACT

9) The midventral mass of erectile tissue in the penis is a:
   A) corpus spongiosum.   B) corpus cavernosum.
   C) spongy urethra.   D) tunica albuginea.

   Answer: A
   Type: MC   Page Ref: 782
   Topic: PENIS

10) The bulb of the penis is formed by the proximal ends of the corpora cavernosa. It is attached to the inferior ramus of the pubis.
    A) Both statements are true.
    B) Both statements are false.
    C) The first statement is true; the second is false.
    D) The second statement is true; the first is false.

    Answer: B
    Type: MC   Page Ref: 782
    Topic: PENIS

11) The tunica albuginea of an ovary:
A) is a layer of loose connective tissue that divides the ovaries into lobes.
B) is the outermost covering of the ovary.
C) is the tissue that contains and surrounds a developing oocyte.
D) is a tissue layer located next to the germinal epithelium.

Answer: D
Type: MC   Page Ref: 784
Topic: OVARIES

12) The corpus luteum:
A) develops from a corpus albicans.
B) develops prior to ovulation.
C) is the zone of the ovary that contains ovarian follicles.
D) none of the above.

Answer: D
Type: MC   Page Ref: 784
Topic: OVARIES

13) The uterine tubes are:
1. the site of zygote formation
2. also called oviducts
3. lined by cells that have microvilli called fimbriae
4. consist of two-thirds ampulla and one-third isthmus
5. attach the ovaries firmly to the uterus
A) 1, 2, 3, 4   B) 1, 3, 5   C) 2, 4, 5   D) 1, 2, 4

Answer: D
Type: MC   Page Ref: 790–791
Topic: UTERINE TUBES

14) Which of the following is *not* a type of ligament that helps maintain the position of the uterus?
A) suspensory ligament   B) round ligament
C) cardinal ligament   D) broad ligament

Answer: A
Type: MC   Page Ref: 791
Topic: UTERUS

15) Which of the following is *not* true of the stratum basalis?
A) It is the permanent layer of the endometrium and it gives rise monthly to the stratum functionalis.
B) It is supplied with blood by the spiral arterioles.
C) It is the layer of the endometrium closest to the myometrium.
D) None of the above; i.e., all are true.

Answer: B
Type: MC   Page Ref: 793
Topic: UTERUS

16) The pudendum:
A) is the recess in the superior portion of the vagina that surrounds the cervix.
B) is the female external genitalia.
C) consists of an anterior urogenital triangle and a posterior anal triangle.
D) is a ligament of the uterus.

Answer: B
Type: MC   Page Ref: 795
Topic: VULVA

17) The onset of menopause occurs when:
A) primary follicles become less responsive to FSH and LH.
B) GnRH release patterns change.
C) estrogen and progesterone levels decline.
D) all of the above.

Answer: D
Type: MC   Page Ref: 801, 803
Topic: AGING AND THE REPRODUCTIVE SYSTEM

TRUE/FALSE. Write 'T' if the statement is true and 'F' if the statement is false.

1) The scrotum provides an environment for the testes that is normally three degrees higher than normal core body temperature.

Answer: FALSE
Type: TF   Page Ref: 771
Topic: SCROTUM

2) Failure of the testes to descend is a condition called cryptorchidism.

Answer: TRUE
Type: TF   Page Ref: 771
Topic: TESTES

3) Tight junctions between sustentacular cells prevent an immune response against developing spermatozoa.

Answer: TRUE
Type: TF   Page Ref: 773
Topic: TESTES

4) Spermatogonia are diploid stem cells that are dormant until puberty, when they undergo mitosis and differentiate into diploid primary spermatocytes.

Answer: TRUE
Type: TF   Page Ref: 776
Topic: SPERMATOGENESIS

5) The second nuclear division of meiosis is reduction division.

Answer: FALSE
Type: TF   Page Ref: 776
Topic: SPERMATOGENESIS

6) The inferior portion of the epididymis, from which the vas deferens emerges, is called the head.

Answer: FALSE
Type: TF   Page Ref: 778
Topic: DUCTS OF THE MALE REPRODUCTIVE TRACT

7) Fructose, which is an energy source for sperm, is a component of the secretions of the prostate gland.

Answer: FALSE
Type: TF   Page Ref: 781
Topic: ACCESSORY SEX GLANDS

8) The ovarian ligament anchors the ovaries to the uterus.

Answer: TRUE
Type: TF   Page Ref: 784
Topic: OVARIES

9) Granulosa cells are epithelial cells that surround a secondary oocyte in the ovary.

Answer: TRUE
Type: TF   Page Ref: 784
Topic: OVARIES

10) Peristaltic contractions assist the movement of the ovum into the uterus.

Answer: TRUE
Type: TF   Page Ref: 791
Topic: UTERINE TUBES

11) The three layers of the wall of the uterus are endometrium, myometrium, and stratum basalis.

Answer: FALSE
Type: TF   Page Ref: 792
Topic: UTERUS

12) Ejection of milk occurs in response to the presence of the hormone oxytocin.

Answer: TRUE
Type: TF   Page Ref: 796
Topic: MAMMARY GLANDS

13) A corpus hemorrhagicum develops into a corpus luteum under the influence of LH.

Answer: TRUE
Type: TF   Page Ref: 800
Topic: UTERINE AND OVARIAN CYCLES

14) The embryonic mesonephric ducts develop into the ducts of the male reproductive system and the uterine tubes of the female reproductive system.

Answer: FALSE
Type: TF   Page Ref: 801
Topic: DEVELOPMENTAL ANATOMY

15) SRY is a gene of the X chromosome that controls development of gonads and external genitalia.

Answer: FALSE
Type: TF   Page Ref: 801
Topic: DEVELOPMENTAL ANATOMY

ESSAY.   Write your answer in the space provided or on a separate sheet of paper.

1) Describe the location and the functions of sustentacular cells.

Answer: The sustentacular (Sertoli) cells are large cells interspersed among the spermatogenic cells of the seminiferous tubules. A single cell will extend from the basement membrane (where it is joined to neighboring sustentacular cells by tight junctions) to the lumen of the tubule. The functions include: formation of the blood–testis barrier; phagocytosis of the excess cytoplasm from spermatogenesis; nourishment and support of spermatocytes, spermatids, and sperm; mediation of the effects of testosterone and FSH; and secretion of the hormone inhibin, which helps control the process of spermatogenesis by inhibiting the secretion of FSH by the pituitary.

Type: ES   Page Ref: 773
Topic: TESTES

2) Explain the terms spermatogenesis, spermiogenesis, and spermiation.

Answer: Spermatogenesis refers to the entire process of sperm production, including the mitotic division of spermatogonia, the meiotic divisions that produce haploid gametes, and their development into mature spermatozoa. Spermiogenesis, the final stage of spermatogenesis, is the maturation of spermatids into spermatozoa.
Spermiation is the release of spermatozoa from sustentacular cells into the lumen of a seminiferous tubule.

Type: ES   Page Ref: 776, 777
Topic: SPERMATOGENESIS

3) The liquid portion of semen is secreted by accessory sex glands. Name and describe them and state the contents and functions of their secretions.

Answer: The seminal vesicles, prostate gland, and bulbourethral glands are described on pp. 780–782.
Type: ES   Page Ref: 780–782
Topic: ACCESSORY SEX GLANDS

4) Describe the gross anatomy of the uterus and accompany your answer with a labeled diagram.

Answer: The description and diagram should include: fundus, body, cervix, internal os, external os, cervical canal, uterine cavity, isthmus of uterus. See Fig. 26.12.
Type: ES   Page Ref: 791
Topic: UTERUS

5) Describe the structure of a mammary gland, using the correct terminology for the tissues and ducts.

Answer: See Fig. 26.21. Answer should correctly use the following terms: modified sudoriferous glands, adipose tissue, lobes, lobules, connective tissue, alveoli, secondary tubules, mammary ducts, lactiferous sinuses, lactiferous ducts, nipple, areola, suspensory ligaments of the breast.
Type: ES   Page Ref: 796-798
Topic: MAMMARY GLANDS

6) Describe the events in the uterus during the preovulatory and postovulatory phases of one menstrual cycle.

Answer: 1. Preovulatory phase: Endometrial cells of the stratum basalis, under the influence of estrogen from developing ovarian follicles, divide to produce a new stratum functionalis layer. Endometrial glands and spiral arterioles develop as the stratum functionalis thickens.
2. Postovulatory phase: Due to the influence of estrogen and progesterone from the corpus luteum, the endometrial glands grow and begin secretion of glycogen, and the endometrium becomes thicker and more vascularized.
Type: ES   Page Ref: 800
Topic: UTERINE AND OVARIAN CYCLES

SHORT ANSWER. Write the word or phrase that best completes each statement or answers the question.

1) The band of skeletal muscle in the spermatic cord that contracts in response to cold temperatures and thereby elevates the testes is called the _____ muscle.

Answer: cremaster
Type: SA   Page Ref: 771
Topic: SCROTUM

2) Each testis is divided into lobules by extensions of the tunica _____.

Answer: albuginea
Type: SA   Page Ref: 772
Topic: TESTES

3) Testosterone is secreted by cells in the testes called _____.

Answer: interstitial endocrinocytes or interstitial cells of Leydig
Type: SA   Page Ref: 773
Topic: TESTES

4) _____ cells of the testes secrete the hormone inhibin and fluid for sperm transport.

Answer: sustentacular
Type: SA   Page Ref: 773
Topic: TESTES

5) In the process of spermatogenesis, meiosis II produces cells called _____, and they contain _____ chromosomes.

Answer: spermatids, 23
Type: SA   Page Ref: 776
Topic: SPERMATOGENESIS

6)   As a spermatid develops into a spermatozoon, it develops a head with a lysosome-like, enzyme-containing granule called a/an _____, and a tail which actually is a/an _____.

Answer: acrosome, flagellum
Type: SA   Page Ref: 776
Topic: SPERMATOGENESIS

7)   The oblique passageway in the anterior abdominal wall that contains the spermatic cord in the male and the round ligament in the female is known as the _____ canal.

Answer: inguinal
Type: SA   Page Ref: 779
Topic: DUCTS OF THE MALE REPRODUCTIVE TRACT

8)   The ejaculatory duct is formed by the union of the vas deferens and the _____.

Answer: duct of the seminal vesicle
Type: SA   Page Ref: 779
Topic: DUCTS OF THE MALE REPRODUCTIVE TRACT

9)   The longest of the three portions of the male urethra is the _____ urethra.

Answer: spongy (penile)
Type: SA   Page Ref: 780
Topic: DUCTS OF THE MALE REPRODUCTIVE TRACT

10) The funnel-shaped distal end of each uterine tube is called a/an _____.

Answer: infundibulum
Type: SA   Page Ref: 790
Topic: UTERINE TUBES

11) Normally the uterus is in a/an _____ position in which the body of the uterus is anterior and superior to the urinary bladder, and the cervix enters the anterior vaginal wall at almost a right angle.

Answer: anteflexed
Type: SA   Page Ref: 791
Topic: UTERUS

12) The layer of the endometrium known as the _____ is permanent, and gives rise to a new _____ layer after each menstruation.

Answer: stratum basalis, stratum functionalis
Type: SA   Page Ref: 792
Topic: UTERUS

13) The female reproductive organ that bears rugae on its inner surface is the _____.

Answer: vagina
Type: SA   Page Ref: 794
Topic: VAGINA

14) The paraurethral glands and the greater vestibular glands secrete _____.

Answer: mucus
Type: SA   Page Ref: 794
Topic: VULVA

MATCHING. Choose the item in Column 2 that best matches each item in Column 1.

Match the stages of spermatogenesis in Column 1 with their descriptions in Column 2.

1)   Column 1: primordial germ cell
         Column 2: found only in the
                 embryonic testes

Answer: found only in the embryonic testes
Type: MA   Page Ref: 776
Topic: SPERMATOGENESIS

2)   Column 1: spermatogonium
         Column 2: develops in embryonic
                 testes; remains dormant
                 until puberty

Answer: develops in embryonic testes; remains dormant until puberty
Type: MA   Page Ref: 776
Topic: SPERMATOGENESIS

3)   Column 1: spermatogonium
         Column 2: differentiates into primary
                 spermatocytes

Answer: differentiates into primary spermatocytes
Type: MA   Page Ref: 776
Topic: SPERMATOGENESIS

4)   Column 1: primary spermatocyte
         Column 2: cell that undergoes
                 reduction division
                 (meiosis I)

Answer: cell that undergoes reduction division
         (meiosis I)
Type: MA   Page Ref: 776
Topic: SPERMATOGENESIS

5)   Column 1: secondary spermatocyte
         Column 2: haploid cell containing 23
                 chromosomes that each
                 consist of 2 chromatids

Answer: haploid cell containing 23 chromosomes that each consist of 2 chromatids
Type: MA   Page Ref: 778
Topic: SPERMATOGENESIS

6)   Column 1: secondary spermatocyte
         Column 2: cell that undergoes
                 equatorial division
                 (meiosis II)

Answer: cell that undergoes equatorial division
         (meiosis II)
Type: MA   Page Ref: 778
Topic: SPERMATOGENESIS

7)   Column 1: spermatid
      Column 2: haploid cell that results
               from the completion of
               meiosis

      Answer: haploid cell that results from the completion of meiosis
      Type: MA   Page Ref: 778
      Topic: SPERMATOGENESIS

Trace the route traveled by a spermatozoon from the time it is released from a sustentacular cell until it leaves the body. Match the ducts in Column 2 with the correct number in Column 1.

8)   Column 1: 1
      Column 2: seminiferous tubule

      Answer: seminiferous tubule
      Type: MA   Page Ref: 777
      Topic: DUCTS OF THE MALE REPRODUCTIVE TRACT

9)   Column 1: 2
      Column 2: straight tubule

      Answer: straight tubule
      Type: MA   Page Ref: 777
      Topic: DUCTS OF THE MALE REPRODUCTIVE TRACT

10)   Column 1: 3
      Column 2: rete testis

      Answer: rete testis
      Type: MA   Page Ref: 777
      Topic: DUCTS OF THE MALE REPRODUCTIVE TRACT

11)   Column 1: 4
      Column 2: efferent duct

      Answer: efferent duct
      Type: MA   Page Ref: 777
      Topic: DUCTS OF THE MALE REPRODUCTIVE TRACT

12)   Column 1: 5
      Column 2: ductus epididymis

      Answer: ductus epididymis
      Type: MA   Page Ref: 777
      Topic: DUCTS OF THE MALE REPRODUCTIVE TRACT

13)   Column 1: 6
      Column 2: ductus deferens

      Answer: ductus deferens
      Type: MA   Page Ref: 778
      Topic: DUCTS OF THE MALE REPRODUCTIVE TRACT

14) Column 1: 7
Column 2: ejaculatory duct

Answer: ejaculatory duct
Type: MA   Page Ref: 779
Topic: DUCTS OF THE MALE REPRODUCTIVE TRACT

15) Column 1: 8
Column 2: prostatic urethra

Answer: prostatic urethra
Type: MA   Page Ref: 780
Topic: DUCTS OF THE MALE REPRODUCTIVE TRACT

16) Column 1: 9
Column 2: membranous urethra

Answer: membranous urethra
Type: MA   Page Ref: 780
Topic: DUCTS OF THE MALE REPRODUCTIVE TRACT

17) Column 1: 10
Column 2: spongy urethra

Answer: spongy urethra
Type: MA   Page Ref: 780
Topic: DUCTS OF THE MALE REPRODUCTIVE TRACT

Match the cells and associated structures of oogenesis in Column 1 with the descriptions in Column 2.

18) Column 1: oogonium
Column 2: develops from a primordial
germ cell during early fetal
development

Answer: develops from a primordial germ cell during early fetal development
Type: MA   Page Ref: 786
Topic: OOGENESIS

19) Column 1: primary oocyte
Column 2: enters meiosis, but remains
in prophase of reduction
division until puberty

Answer: enters meiosis, but remains in prophase of reduction division until puberty
Type: MA   Page Ref: 786
Topic: OOGENESIS

20) Column 1: primordial follicle
Column 2: spherical structure
containing a primary oocyte

Answer: spherical structure containing a primary oocyte
Type: MA   Page Ref: 786
Topic: OOGENESIS

21) Column 1: secondary oocyte
Column 2: proceeds to metaphase of
equatorial division

Answer: proceeds to metaphase of equatorial division
Type: MA   Page Ref: 790
Topic: OOGENESIS

22) Column 1: ovum
Column 2: haploid cell

Answer: haploid cell
Type: MA   Page Ref: 790
Topic: OOGENESIS

23) Column 1: zona pellucida
Column 2: noncellular glycoprotein
layer surrounding the
primary or secondary oocyte

Answer: noncellular glycoprotein layer surrounding the primary or secondary oocyte
Type: MA   Page Ref: 790
Topic: OOGENESIS

24) Column 1: corona radiata
Column 2: inner layer of granulosa
cells

Answer: inner layer of granulosa cells
Type: MA   Page Ref: 790
Topic: OOGENESIS

25) Column 1: antrum
Column 2: fluid–filled cavity of a
secondary follicle

Answer: fluid–filled cavity of a secondary follicle
Type: MA   Page Ref: 790
Topic: OOGENESIS

Match the male structures in Column 1 with their female homologues in Column 2.

26) Column 1: testes
Column 2: ovaries
Foil: uterus

Answer: ovaries
Type: MA   Page Ref: 796
Topic: OVARIES

27) Column 1: clitoris
Column 2: glans penis

Answer: glans penis
Type: MA   Page Ref: 796
Topic: VULVA

28) Column 1: prostate gland
Column 2: paraurethral glands

Answer: paraurethral glands
Type: MA   Page Ref: 796
Topic: VULVA

29) Column 1: bulbourethral glands
Column 2: greater vestibular glands

Answer: greater vestibular glands
Type: MA   Page Ref: 796
Topic: VULVA

30) Column 1: scrotum
Column 2: labia majora

Answer: labia majora
Type: MA   Page Ref: 796
Topic: VULVA

31) Column 1: spongy urethra
Column 2: labia minora

Answer: labia minora
Type: MA   Page Ref: 796
Topic: VULVA

CHAPTER 27   Developmental Anatomy

MULTIPLE CHOICE. Choose the one alternative that best completes the statement or answers the question.

1)   There is a three-day fertile period extending from two days before ovulation to one day after, because:
A) sperm can remain viable for up to three days in the female reproductive tract.
B) a secondary oocyte is viable for about 24 hours after ovulation and sperm are viable for about 48 hours in the female reproductive tract.
C) it takes three days for the sperm to reach the secondary oocyte.
D) it takes three days for the secondary oocyte to reach the uterus.

Answer: B
Type: MC   Page Ref: 811
Topic: FERTILIZATION

2)   The normal site of fertilization is the:
A) uterine tube.   B) infundibulum.   C) uterine cavity.   D) cervical canal.

Answer: A
Type: MC   Page Ref: 811
Topic: FERTILIZATION

3)   Which of the following muscular actions normally assist the travels of the secondary oocyte and/or sperm so that fertilization can occur?
1. oocyte moved by peristalsis of uterine tube
2. sperm moved by peristalsis of uterine tube
3. oocyte moved by uterine contractions
4. sperm moved by uterine contractions
A) 1, 2, 4   B) 2, 4   C) 2, 3   D) 1, 4

Answer: D
Type: MC   Page Ref: 811
Topic: FERTILIZATION

4)   Place the following in the correct order:
1. spermatozoon penetrates corona radiata
2. acrosome releases enzymes
3. syngamy occurs
4. spermatozoon binds to receptors in zona pellucida
5. segmentation nucleus is formed
A) 1, 2, 3, 4, 5   B) 2, 1, 4, 3, 5   C) 4, 2, 1, 5, 3   D) 2, 4, 1, 5, 3

Answer: B
Type: MC   Page Ref: 811
Topic: FERTILIZATION

5) Dizygotic twins result from:
    A) fertilization of two secondary oocytes by different sperm.
    B) fertilization of one secondary oocyte by two sperm.
    C) early splitting of a zygote to produce two individuals.
    D) none of the above.

    Answer: A
    Type: MC   Page Ref: 811
    Topic: FERTILIZATION

6) Calcium ions are released as soon as a spermatozoon enters the secondary oocyte.
    This leads to changes in the secondary oocyte that block the entry of other sperm.
    A) Both statements are true.
    B) Both statements are false.
    C) The first statement is true; the second is false.
    D) The second statement is true; the first is false.

    Answer: A
    Type: MC   Page Ref: 811
    Topic: FERTILIZATION

7) Cleavage:
    A) is completed in the first two hours following fertilization.
    B) increases the number of cells of the embryo.
    C) increases the size of the embryo.
    D) all of the above.

    Answer: B
    Type: MC   Page Ref: 811
    Topic: CLEAVAGE

8) The morula:
    A) is surrounded by the zona pellucida.
    B) when first formed, is the same size as the original zygote.
    C) is a solid mass of cells.
    D) all of the above.

    Answer: D
    Type: MC   Page Ref: 812
    Topic: CLEAVAGE

9) Following implantation, the trophoblast develops into two layers, syncytiotrophoblast
    (an extracellular material) and cytotrophoblast (a cellular layer). The cytotrophoblast
    secretes enzymes that assist the embryo in penetrating the endometrium.
    A) Both statements are true.
    B) Both statements are false.
    C) The first statement is true; the second is false.
    D) The second statement is true; the first is false.

    Answer: B
    Type: MC   Page Ref: 813
    Topic: IMPLANTATION

10) The period of development called the embryonic period is _____ in length.
    A) 2 weeks   B) 1 month   C) 2 months   D) 3 months
    Answer: C
    Type: MC   Page Ref: 815
    Topic: EMBRYONIC DEVELOPMENT

11) The three primary germ layers develop from the _____ of the blastocyst.
    A) syncytiotrophoblast   B) cytotrophoblast
    C) blastocele   D) inner cell mass
    Answer: D
    Type: MC   Page Ref: 815
    Topic: EMBRYONIC DEVELOPMENT

12) The three major events of the embryonic period are:
    A) fertilization, implantation, and differentiation.
    B) implantation, gastrulation, and differentiation.
    C) development of embryonic tissues, embryonic membranes, and placenta.
    D) development of embryonic tissues, embryonic membranes, and organs.
    Answer: C
    Type: MC   Page Ref: 815-819
    Topic: EMBRYONIC DEVELOPMENT

13) The placenta allows which of the following to pass into fetal blood?
    1. maternal blood cells
    2. most bacteria
    3. drugs
    4. AIDS, german measles, and polio viruses
    5. alcohol
    A) 1, 3, 4, 5   B) 3, 4, 5   C) 1, 2, 4, 5   D) 1, 2
    Answer: B
    Type: MC   Page Ref: 819
    Topic: EMBRYONIC DEVELOPMENT

14) Following implantation, the stratum functionalis is modified to become the:
    A) decidua capsularis.   B) decidua parietalis.
    C) decidua basalis.   D) all of the above.
    Answer: D
    Type: MC   Page Ref: 819
    Topic: EMBRYONIC DEVELOPMENT

15) As the chorionic villi grow into the endometrium, they contain blood vessels of the:
    A) decidua.   B) amnion.   C) allantois.   D) yolk sac.
    Answer: C
    Type: MC   Page Ref: 819
    Topic: EMBRYONIC DEVELOPMENT

16) Which of the following would be an expected change in the mother during pregnancy?
A) decreased blood volume
B) decrease in heart rate
C) decreased urination
D) decrease in motility of the gastrointestinal tract

Answer: D
Type: MC   Page Ref: 823
Topic: CHANGES DURING PREGNANCY

17) The hormone _____ causes increased flexibility in certain joints such as the pubic symphysis in the latter part of pregnancy.
A) estrogen   B) relaxin   C) hCG   D) progesterone

Answer: B
Type: MC   Page Ref: 823
Topic: CHANGES DURING PREGNANCY

TRUE/FALSE. Write 'T' if the statement is true and 'F' if the statement is false.

1) In order for fertilization to occur, sperm must be present in the female sometime during the first 24 hours following ovulation.

Answer: TRUE
Type: TF   Page Ref: 811
Topic: FERTILIZATION

2) Syngamy is the fusion of the haploid male and female pronuclei to produce a diploid segmentation nucleus.

Answer: FALSE
Type: TF   Page Ref: 811
Topic: FERTILIZATION

3) The morula develops into a blastocyst by day five, at which time the embryo enters the uterine cavity.

Answer: TRUE
Type: TF   Page Ref: 812
Topic: CLEAVAGE

4) The three main parts of the blastocyst are trophoblast, inner cell mass, and blastocele.

Answer: TRUE
Type: TF   Page Ref: 812
Topic: CLEAVAGE

5) The umbilical cord contains three blood vessels: two umbilical arteries and one umbilical vein.

Answer: TRUE
Type: TF   Page Ref: 821
Topic: EMBRYONIC DEVELOPMENT

6) The embryonic disc appears first as a bilayered structure of ectoderm and endoderm next to the amniotic cavity.

Answer: TRUE
Type: TF   Page Ref: 817
Topic: EMBRYONIC DEVELOPMENT

7) The ectoderm of the embryonic disc eventually gives rise to all nervous tissue, the epidermis of the skin, and epithelium of the oral and nasal cavities.

Answer: TRUE
Type: TF   Page Ref: 818
Topic: EMBRYONIC DEVELOPMENT

8) One advantage of chorionic villi sampling over amniocentesis is that it can be done 6-8 weeks earlier than amniocentesis.

Answer: TRUE
Type: TF   Page Ref: 826
Topic: PRENATAL DIAGNOSTIC TESTS

9) One advantage of amniocentesis over chorionic villi sampling is that amniocentesis has a lower probability of causing a spontaneous abortion.

Answer: TRUE
Type: TF   Page Ref: 827
Topic: PRENATAL DIAGNOSTIC TESTS

ESSAY. Write your answer in the space provided or on a separate sheet of paper.

1) Describe the production and functions of the amniotic fluid.

Answer: It is initially formed as a filtrate of plasma from maternal blood, but later contains fetal urine and sloughed-off embryonic cells. It serves as a shock absorber, temperature regulator, and prevents adhesion of the developing fetus to surrounding tissues.
Type: ES   Page Ref: 817-818
Topic: EMBRYONIC DEVELOPMENT

2) List the stages of labor and briefly describe the events that occur in each stage.

Answer: 1. Stage of dilation
2. Stage of expulsion of fetus
3. Stage of expulsion of placenta
Descriptions are on pp. 825.
Type: ES   Page Ref: 825
Topic: LABOR

3) Following the deposition of the sperm in the female reproductive tract, describe the events leading up to and including fertilization.

Answer: The answer should include a description of capacitation of sperm and events leading up to the formation of the segmentation nucleus of the zygote, as found on pp. 811.
Type: ES   Page Ref: 811
Topic: FERTILIZATION

4)   Place the following events in correct order and briefly define each: implantation, fertilization, fetal period, labor, cleavage, embryonic development, placental development.

Answer: 1. Fertilization: sperm cell enters a secondary oocyte; nuclear material merges into a single nucleus of the zygote
2. Cleavage: early divisions of the zygote
3. Implantation: attachment of the blastocyst to the endometrium
4. Embryonic development: first two months of development during which the rudiments of organs and the embryonic membranes develop
5. Placental development: production of the placenta from the embryonic chorion and the maternal decidua basalis, by the end of the third month of development
6. Fetal period: period of growth from second month to birth
7. Labor: process of expulsion of fetus from the uterus and through the vagina

Type: ES   Page Ref: 811–823
Topic: STAGES OF DEVELOPMENT

SHORT ANSWER.   Write the word or phrase that best completes each statement or answers the question.

1)   A zygote consists of a fertilized ovum and _____.

Answer: zona pellucida
Type: SA   Page Ref: 811
Topic: FERTILIZATION

2)   The fusion of the male pronucleus with the female pronucleus produces a _____ nucleus.

Answer: segmentation
Type: SA   Page Ref: 811
Topic: FERTILIZATION

3)   The cells of the embryo produced by cleavage are called _____.

Answer: blastomeres
Type: SA   Page Ref: 812
Topic: CLEAVAGE

4)   The functional changes in sperm while in the female reproductive tract, preparing the sperm to fertilize a secondary oocyte, are collectively referred to as _____.

Answer: capacitation
Type: SA   Page Ref: 811
Topic: FERTILIZATION

5)   Implantation usually occurs on the _____ wall of the fundus or body of the uterus.

Answer: posterior
Type: SA   Page Ref: 813
Topic: IMPLANTATION

6) Differentiation of the tissue of the embryo into the three primary germ layers begins _____ (before or after) implantation.

Answer: after
Type: SA   Page Ref: 815
Topic: EMBRYONIC DEVELOPMENT

7) The embryonic layer called the _____ develops from the trophoblast of the blastocyst.

Answer: chorion
Type: SA   Page Ref: 819
Topic: EMBRYONIC DEVELOPMENT

8) The portion of the decidua that separates the embryo from the uterine cavity is called the _____.

Answer: decidua capsularis
Type: SA   Page Ref: 819
Topic: EMBRYONIC DEVELOPMENT

9) The maternal part of the placenta is formed from the decidua _____.

Answer: basalis
Type: SA   Page Ref: 819
Topic: EMBRYONIC DEVELOPMENT

10) Deoxygenated, waste-laden blood leaves the fetus through a blood vessel called an umbilical _____.

Answer: artery
Type: SA   Page Ref: 819
Topic: EMBRYONIC DEVELOPMENT

11) In the placenta, nutrients, gases, and wastes are exchanged between blood vessels of the chorionic villi and _____ of the maternal portion of the placenta.

Answer: intervillous spaces or sinuses
Type: SA   Page Ref: 819
Topic: EMBRYONIC DEVELOPMENT

12) In the months of development after the _____ month, the embryo is referred to as a fetus.

Answer: second
Type: SA   Page Ref: 821
Topic: FETAL GROWTH

13) A procedure done at 14 to 16 weeks of gestation to test for the presence of certain genetic conditions such as spina bifida and hemophilia is called _____.

Answer: amniocentesis
Type: SA   Page Ref: 826
Topic: PRENATAL DIAGNOSTIC TESTS

14) The six-week period following delivery, in which maternal organs and physiology return to the prepregnancy state, is called the _____.

Answer: puerperium
Type: SA   Page Ref: 826
Topic: LABOR

MATCHING. Choose the item in Column 2 that best matches each item in Column 1.

Match the stages of development in Column 1 with their descriptions in Column 2.

1)   Column 1: morula
        Column 2: solid ball of cells resulting
                from cleavage

        Answer: solid ball of cells resulting from cleavage
        Type: MA   Page Ref: 811
        Topic: CLEAVAGE

2)   Column 1: zygote
        Column 2: a fertilized ovum,
                containing a segmentation
                nucleus and surrounded by
                the zona pellucida

        Answer: a fertilized ovum, containing a segmentation nucleus and surrounded by the
                zona pellucida
        Type: MA   Page Ref: 811
        Topic: FERTILIZATION

3)   Column 1: blastocyst
        Column 2: hollow ball of cells

        Answer: hollow ball of cells
        Type: MA   Page Ref: 812
        Topic: CLEAVAGE

4)   Column 1: blastocyst
        Column 2: stage of development that
                enters the uterus

        Answer: stage of development that enters the uterus
        Type: MA   Page Ref: 812
        Topic: CLEAVAGE

5)   Column 1: blastomere
        Column 2: small cells produced by the
                cell divisions during
                cleavage

        Answer: small cells produced by the cell divisions during cleavage
        Type: MA   Page Ref: 812
        Topic: CLEAVAGE

6) Column 1: trophoblast
    Column 2: outer covering of cells of
              the blastocyst

    Answer: outer covering of cells of the blastocyst
    Type: MA   Page Ref: 812
    Topic: CLEAVAGE

7) Column 1: embryoblast
    Column 2: inner cell mass that
              develops into the embryo

    Answer: inner cell mass that develops into the embryo
    Type: MA   Page Ref: 812
    Topic: CLEAVAGE

Match the events of fetal development in Column 2 with the numbers in Column 1 that indicate the end of the month in which the events occur.

8) Column 1: 1
    Column 2: heart forms and begins to
              beat

    Answer: heart forms and begins to beat
    Type: MA   Page Ref: 822
    Topic: FETAL GROWTH

9) Column 1: 2
    Column 2: ossification begins

    Answer: ossification begins
    Type: MA   Page Ref: 822
    Topic: FETAL GROWTH

10) Column 1: 3
    Column 2: heartbeat can be detected

    Answer: heartbeat can be detected
    Type: MA   Page Ref: 822
    Topic: FETAL GROWTH

11) Column 1: 4
    Column 2: hair appears on head and
              articulations begin to form

    Answer: hair appears on head and articulations begin to form
    Type: MA   Page Ref: 822
    Topic: FETAL GROWTH

12) Column 1: 5
    Column 2: fetal movements are first
              felt by mother

    Answer: fetal movements are first felt by mother
    Type: MA   Page Ref: 822
    Topic: FETAL GROWTH

13) Column 1: 6
   Column 2: type II alveolar cells begin
             to produce surfactant

   Answer: type II alveolar cells begin to produce surfactant
   Type: MA   Page Ref: 822
   Topic: FETAL GROWTH

14) Column 1: 7
   Column 2: fetus is capable of survival
             if premature birth occurs at
             this stage

   Answer: fetus is capable of survival if premature birth occurs at this stage
   Type: MA   Page Ref: 822
   Topic: FETAL GROWTH

15) Column 1: 8
   Column 2: testes of male fetus descend
             into scrotum

   Answer: testes of male fetus descend into scrotum
   Type: MA   Page Ref: 822
   Topic: FETAL GROWTH

16) Column 1: 9
   Column 2: lanugo is shed

   Answer: lanugo is shed
   Type: MA   Page Ref: 822
   Topic: FETAL GROWTH